THIRD EDITION

Great Ideas for Teaching Astronomy

Stephen M. Pompea
Steward Observatory, University of Arizona

Australia • Canada • Mexico • Singapore
Spain • United Kingdom • United States

Assistant Editor: *Marie Carigma-Sambilay*
Marketing Manager: *Steve Catalano*
Editorial Assistant: *Samuel Subity*
Production: *Dorothy Bell*
Cover Design: *Vernon T. Boes*
Cover Image: *NASA and The Hubble Heritage Team (STScI)*
Printing and Binding: *Webcom Limited*

For more information about this or any other Brooks/Cole product, contact:
BROOKS/COLE
511 Forest Lodge Road
Pacific Grove, CA 93950 USA
www.brookscole.com
1-800-423-0563 (Thomson Learning Academic Resource Center)

ABOUT THE COVER: In the direction of the constellation Canis Major, two spiral galaxies pass by each other like majestic ships in the night. The near-collision has been caught in images taken by NASA's Hubble Space Telescope and its Wide Field Planetary Camera 2.

The larger and more massive galaxy is cataloged as NGC 2207 (on the left in the Hubble Heritage image), and the smaller one on the right is IC 2163. Strong tidal forces from NGC 2207 have distorted the shape of IC 2163, flinging out stars and gas into long streamers stretching out a hundred thousand light-years toward the right-hand edge of the image.

Trapped in their mutual orbit around each other, these two galaxies will continue to distort and disrupt each other. Eventually, billions of years from now, they will merge into a single, more massive galaxy. It is believed that many present-day galaxies, including the Milky Way, were assembled from a similar process of coalescence of smaller galaxies occurring over billions of years.

Printed in Canada

10 9 8 7 6 5 4 3 2 1

ISBN 0-534-37301-1

About the Editor

Stephen M. Pompea received his bachelors degree in physics, space physics, and astronomy from Rice University, his masters in physics teaching from Colorado State University (where he was an adjunct faculty member for ten years), and his doctorate degree in astronomy from the University of Arizona. He has served as an Associate Scientist on the Gemini 8-Meter Telescopes Project and as Instrument Scientist for the NICMOS (Near Infrared Camera and Multi-Object Spectrometer) Project for the Hubble Space Telescope. In addition to his technical and educational consulting, he is an Adjunct Associate Astronomer at Steward Observatory, University of Arizona.

His major research interests include star formation processes in galaxies, development of ultrablack surfaces for space telescopes, adaptive and active optics, telescope and instrument design, and stray light and contamination issues in optical systems. His interests in science education are wide ranging and involve program design in both formal and informal education. He has served as an educational consultant to a variety of programs including the NASA Classroom of the Future program at Wheeling Jesuit University, the Sun-Earth Connection Education Forum, the LodeStar Project of New Mexico, and the Space Science Institute. He has given numerous workshops devoted to improving science teaching and to the training of secondary school and college science faculty.

Dr. Pompea is a member of the American Astronomical Society, the Astronomical Society of the Pacific, the National Science Teachers Association, the Optical Society of America, and the Society of Photo-Optical Instrumentation Engineers. He lives in Tucson, Arizona where he conducts research and consults in optical physics and science education. He can be reached at (520) 792-2366 or by email at spompea@as.arizona.edu.

About the Series

The Great Ideas series began in late 1989 with the first edition of this book. The interest sparked in new teaching ideas by the astronomy book stimulated two other books: *Great Ideas for Teaching Physics* in 1991 and *Great Ideas for Teaching Geology in 1992*. With the great success of the Great Ideas series, we recognize the value of making these books available to teachers at all levels on the widest possible basis.

Contents

Introduction

College teachers often receive very little instruction or support in the art of effective teaching. The assumption is that knowledge of the subject matter without training in teaching skills is sufficient for teaching success. Nothing could be further from the truth. The on-the-job "training" graduate students receive when thrust into their first lab, discussion section, or lecture is often not a positive experience conducive to forming good teaching habits. Later, after the basics of teaching survival have been learned, there is often little support for the development of new teaching ideas or approaches. On the other hand, high school and middle school teachers usually have received extensive training in the art of teaching but are often resource limited or are teaching at the edge of their subject matter training. Their support system is also limited, and they find themselves extremely limited in the amount of time they can spend on course development.

This book provides ideas that can be used to improve astronomy teaching. It is also a support vehicle to encourage and assist teachers in trying new approaches. This third edition is enlarged considerably from previous editions and includes ideas that will appeal to teachers at all levels of astronomy teaching. The resource area and the section on astrobiology have been considerably expanded, and new sections on public outreach, research, pedagogy, and the national science education standards have been added. With the impressive advent of the World Wide Web, I have tried to provide some of my favorite web references.

This book does not delve as deeply as I would like into the practical side of teaching and managing a class, even though there is much room for improvement in this area. I have seen brilliant lecturers who were not appreciated by students because their tests were poorly constructed, or because they forgot to give a minimal amount of encouragement to students after a particularly hard test. Such relatively easy-to-correct deficiencies often affect a student's opinion of a class or subject far more than hours of preparation by an instructor. Just as a poor bedside manner may turn a patient against a brilliant doctor, the lack of encouragement, and the dearth of humor and playfulness of many college science teachers often turns students against an intrinsically interesting subject. This book's primary role may be to introduce some more playful ideas or demonstrations into an instructor's repertoire so that both teacher and student may get greater enjoyment from the teaching and learning process. Perhaps this book may help us remember how much fun astronomy is, and is to teach.

I would like to thank Marie Carigma-Sambilay and Gary Carlson at Brooks/Cole for their patience and support of this project and my wife Nancy, a master teacher, for her many contributions to this book.

Stephen M. Pompea
Tucson, Arizona

Contributors

Jeff Adams
Department of Physics
Montana State University
Bozeman, MT 59717-3840
adams@physics.montana.edu

Elizabeth M. Alvarez del Castillo
International Dark-Sky Association
3225 N. First Avenue
Tucson, AZ 85719-2103
liz@darksky.org

Mel Anderson
Pittsburg Middle School
399 W. Leighton
Frontenac, KS 66763
anderson@cjnetworks.com

Bob Bechtel
Department of Physics
Vincennes University
Vincennes IN 47591

David C. Blewett
Butler County Community College
College Drive, Oak Hills
Butler PA 16001

John Broderick
Physics Department
Virginia Polytechnic Institute and State University
Blacksburg VA 24061-0435
john.broderick@vt.edu

Michael Broyles
Collin County Community College
2200 West University
McKinney, TX 75070

David Bruning
University of Louisville
Louisville, KY 40292
Stellar Research & Education
P.O. Box 1223
Waukesha,WI 53187-1223

Buck Buchannon
Belgrade Middle School
Box 166
Triple Crown Rd.
Manhattan, Montana 59741
buck@avicom.net

Greg Burley
Department of Geophysics and Astronomy
University of British Columbia
Vancouver, B.C. V6T 1Z4 Canada

John Burns
Mt. San Antonio Community College District
Walnut, CA 91789

Kimberly Burtnyk
Department of Physics MC 3840
Montana State University
Bozeman, MT 59717
burtnyk@physics.montana.edu

Robert G. Carson
Department of Physics
Rollins College
Winter Park FL 32789-4497

Nahide Craig
Space Sciences Laboratory
University of California, Berkeley
Grizzly Peak @ Centennial Drive
Berkeley, CA 94720-7450
ncraig@ssl.berkeley.edu

Tom Damon
Pikes Peak Community College
Colorado Springs, CO 80906

Edna DeVore
Director of Educational Programs
SETI Institute
2035 Landings Dr.
Mountain View, CA 94043
edevore@seti.org

Janet Erickson
C.R. Anderson Middle School
1200 Knight
Helena, MT 59601
football@ixi.net

Danny R. Faulkner
University of South Carolina at Lancaster
Lancaster, SC 29720

Alex Filippenko
Department of Astronomy
University of California
Berkeley, CA 94720

Karl Giberson
Department of Physics
Eastern Nazarene College
Quincy MA 02170

Alan Gould
Lawrence Hall of Science
University of California, Berkeley
Berkeley, CA 94720
agould@uclink4.berkeley.edu

Donna L. Governor
Brown Barge Middle School
151 E. Fairfield Dr.
Pensacola, FL 32503
dulci@pcola.gulf.net

Isabel Hawkins
Space Sciences Laboratory
University of California, Berkeley
Grizzly Peak @ Centennial Drive
Berkeley, CA 94720-7450
isabelh@ssl.berkeley.edu

Paul Heckert
Department of Chemistry and Physics
Western Carolina University
Cullowhee, NC 28723
heckert@wcu.edu

Thomas Hockey
Department of Earth Science
University of Northern Iowa
Cedar Falls IA 50614
hockey@uni.ed

Darrel Hoff
Department of Physics
Luther College
Decorah IA 52101
hoffdarr@luther.edu

Chris Impey
Steward Observatory
University of Arizona
Tucson, AZ 85721
Impey@as.arizona.edu

James E. Kettler
Department of Physics
Ohio University - Eastern Campus
45425 National Road
St. Clairsville, OH 43950

Eric Kincanon
Gonzaga University
Spokane WA 99258-0001

Karl Kuhn
Department of Physics and Astronomy
Eastern Kentucky University
Richmond KY 40475-0950

John B. Laird
Department of Physics and Astronomy
Bowling Green State University
Bowling Green, OH 43403
laird@tycho.bgsu.edu

V. Gordon Lind
Department of Physics
Utah State University
Logan UT 84322-4415

Joseph W. Mast
Math-Science Department
Eastern Mennonite College
Harrisonburg VA 22801

Gene Maynard
Department of Physical Science
Radford University
Radford, VA 21142

Don McCarthy
Steward Observatory
University of Arizona
933 N. Cherry Ave.
Tucson, AZ 85721-0065

Karen Meech
Institute for Astronomy
University of Hawaii
Honolulu, HI 96822
meech@ifa.hawaii.edu

Michael E. Mickelson
Department of Physics and Astronomy
Denison University
Granville, OH 43023

Eberhard Moebius
Space Science Center
University of New Hampshire
Durham, NH 0382
Eberhard.Moebius@unh.edu

Cherilynn A. Morrow
Space Science Institute
1540 30th St.
Boulder, CO 80303-1012
camorrow@colorado.edu

Paul Nyhuis
Inver Hills Community College
Inver Grove Heights, MN 55076

Danny Overcash
Lenoir-Rhyne College
Hickory, NC 28603

Stephen M. Pompea
Pompea & Associates
1321 East Tenth Street
Tucson AZ 85719-5808
spompea@as.arizona.edu

David B. Porter
Portland Community College
Portland, OR 97219

Edward E. Prather
Montana State University
Department of Physics
Bozeman, MT 59717-3840

Travis A. Rector
National Optical Astronomy
Observatories
Tucson, AZ
rector@noao.edu

Charles Roberts
Department of Physics
S.U.N.Y. at Morrisville
Morrisville NY 13408

Elizabeth Roettger
DePaul University
1 E. Jackson, Chicago, IL 60604
roettger@ix.netcom.com

Joanne Rosvick
Department of Physics and Astronomy
University of Victoria
P.O. Box 3055
Victoria, B.C. V8W 3P7 Canada

Neal P. Rowell
University of South Alabama
Mobile AL 36688

Philip Sakimoto
Whitman College
Walla Walla, WA 99362

Greg Schultz
Space Sciences Laboratory
University of California, Berkeley
Grizzly Peak @ Centennial Drive
Berkeley, CA 94720-7450
schultz@ssl.berkeley.edu

Richard D. Schwartz
Department of Physics and Astronomy
University of Missouri-St. Louis
St. Louis, MO 63121
schwartz@newton.umsl.ed

Sandy Shutey
Butte High School,
401 S. Wyoming
Butte, MT 59701
sshutey@in-tch.com

Alison F. Simmerman
P.O. Box 1837
Honoka'a, HI 96727
asimmerm@k12.hi.us

Tim Slater
Department of Physics
Montana State University
Bozeman, MT 59717-3840
tslater@physics.montana.edu

Glenn M. Spiczak
Bartol Research Institute
University of Delaware
217 Sharp Laboratory
Newark, DE 19716
spiczak@bartol.udel.edu

Ronald Stoner
Department of Physics
Bowling Green State University
Bowling Green, OH 43403

Thomas L. Swihart
Deceased, formerly of
Steward Observatory
University of Arizona
Tucson AZ 85721

Constance E. Walker
Steward Observatory
University of Arizona,
Tucson, AZ 85721
connie@as.arizona.edu

Anthony Weitenbeck
University of Wisconsin Baraboo/Sauk
1006 Connie Rd.
Baraboo WI 53913
ajw@sal.wisc.edu

William G. Weller
Association of Universities for
Research in Astronomy
P.O. Box 26732
Tucson AZ 85726-6732

Marty Wells
LaSalle High School
3091 North Bend Rd.
Cincinnati, OH 45239
mcwells@worldnet.att.net

Carl J. Wenning
Department of Physics
Illinois State University
Normal IL 61761-6901

Dan Wilkins
University of Nebraska - Omaha
Omaha NE 68182

Louis Winkler
Department of Astronomy and
Astrophysics
The Pennsylvania State University
University Park PA 16802

Jeffrey A. Yuhas
Reading Memorial High School
62 Oakland Road
Reading, MA 01867
jayuhas@massed.net

Chapter 1
Connecting with Students

Large Lecture Hall Survival Guidelines

Here are some very basic survival rules for a large lecture course for those (such as graduate students) who are new to this format. The advice I give here has proven successful many times and has been presented to college teachers in a much more detailed way in workshops with great success. Here is the condensed version to get you started.

Variety is extremely important in your lecture style. Some students learn best from hearing about it, others learn best through visual stimulation and analogies, and others really have to touch the material to absorb it. We should try to appeal to verbal, visual, and kinesthetic modalities as often as possible. Demonstration, jokes, cartoons, and a sense of the dramatic provide a way to interest students and also appeal to all of their senses.

Many teachers stop their preparation for a lecture once the basic subject matter is on a viewgraph. This is unfortunate as the fun part is next, and further preparation will increase the positive feedback a teacher receives from students. The next phase of preparation is to ask: "How am I going to keep the students interested and receptive? How am I going to start the class off on the right foot? How am I going to end the class? How am I going to keep their attention?" Here are some approaches to answering these questions. Each class session can be viewed as a speech. Nobody starts a speech (especially to a paying audience, as students really are) by saying "We'll continue talking today about the satellites of Jupiter...." A speech must start with an attention-getting step. Then you must convince the audience that there is a need to listen carefully. Then comes the body of the speech. Students have heard many bad lectures, and need to be convinced right away that this one is different.

What makes a "good" lecture, in the eyes of students? A sense of organization is very important as students do not have a firm mental outline of the subject and are trying very hard to form a mental map of the material. A great deal of student frustration arises from not knowing where to hang things. The most valuable teaching tip I have learned is to put an outline of the lecture on the board, or somewhere where it can be seen all the time. That way a student can track the progress and, if they drift away, can come back. It makes the lecture come in bite-sized chunks, as well. Part of putting up an outline is that it promises students that you will not be wasting their time, that you are a professional at teaching and know what you are doing. How to start a class? Starting each class off with a cartoon (e.g., Sydney Harris) is a good way to begin. A dramatic metaphor for each class might not be bad either.

Once, I filled in for an instructor whose class was studying the distance scale ladder. I first prepared a set of viewgraphs, probably not too different from the sets many of us use for such a lecture. Then I found a cartoon that related to the subject. Then I wondered how I would keep from being bored when talking about this subject, or keep the students from being bored. I thought about the word ladder. I went and borrowed a ladder from the custodian, realizing that each rung on the ladder was a different part of the lecture. For example, I put an apple on one rung of the ladder to symbolize Newton's apple and the law of universal gravitation. It represented using the period of double stars to derive their distance. On another rung I put a flashlight to symbolize the concept of standard candles. On another rung I put a credit card to symbolize the belief that the laws of physics are applicable to other places in the universe, just like MasterCard is good anywhere in the world. The best part was that I put all these items on the ladder before the lecture, as students were coming in, and then went up to retrieve each one as I came to that part of the lecture. The students could hardly wait until I got to the top of this large ladder and fell off. They were getting more and more excited as the class progressed. That is a valuable goal, that the students get more excited as the lecture (or the course) progresses.

How do you end a lecture? In many classrooms, the students grow restless and start to become distant when there are ten minutes left. This is a waste of good time. I am a firm believer in a test at the end of each lecture. If there were certain key concepts or facts that you wanted students to learn, why not ask them if they indeed learned it. A short, 2-minute, 5-short-answer quiz can help students feel confident with what they have learned, and the quiz reinforce, it as well. It is not a bad idea to have them turn them in once in awhile.

I have listened to many excellent lectures (to me at least) delivered by some excellent astronomers. Yet, when I was younger and could pretend that I was a student I often heard grumbling afterwards from students. I puzzled at how a student could not appreciate a well-reasoned and organized lecture. I have come to the following conclusion:

A lecture may be well-organized from the point of view of someone who knows the material, but may be rather disorganized to someone who is just learning the subject. Logical organization is not necessarily meaningful unless the student knows the trend. Students often have no sense of what is important in new subject area. A lecture is sometimes equivalent to having someone read the book to you, only there are no headings, bold and italic type, and the organization is not as good. How can the lecturer hope to compete with a much better-organized book?

The answer is that the lecturer must bring something new to the subject. The only things that can be brought are a sense of the personal and a sense of drama. Astronomy can be taught as a series of burning questions and as a set of golden threads that run through it. It is the instructor's job to introduce new questions, seek answers, to motivate students, and to reassure students. The ideas presented here can serve as ways to spice up a lecture, lend a human side to it, or demonstrate or simulate a natural phenomena. In summary, I advocate a three-stage process for preparing a lecture. First, prepare the subject matter. Second, find a way to make it more interesting and more human, and more attached to different aspects of a student's life. And third, ask how am I going to do something really memorable in this lecture. Being unpredictable is very important in keeping student interest.

One final thought is that teachers need to make more demands on students, and not tolerate poor behavior in class. Take pride in your ability and expect the best from students. I regularly ask that students sit closer to the front where they can participate. For example, you can put a sign on the viewgraph projector announcing that students should only sit in the first 15 rows. If you have a cordless microphone, walk around and keep (and demand) their attention.

Stephen M. Pompea
Pompea & Associates

Attracting and Keeping the Attention of Students I

Just before the very first lecture in a large introductory astronomy class, I set up a carousel of slides of assorted astronomical objects. The slides are projected onto a large screen at the front of the class, and they advance automatically (say, once every 6 seconds) while students are walking in and taking their seats. Cosmic-sounding music (e.g., Pink Floyd) fills the room. This sets the mood for the course: students are awed and intrigued by the photographs, and they are likely to pay closer attention to what I say. Moreover, students who are hanging around and thinking of auditing the course are more likely to be attracted to it. I also play music and show slides before a few other lectures throughout the semester; this helps maintain interest by reminding the students how beautiful the universe is.

Prior to the midterm examinations, as students are walking into the room and taking their seats, I show slides of some of my vacations in exotic places. This helps put the students at ease.

Alex Filippenko
University of California, Berkeley

Attracting and Keeping the Attention of Students II

When lecturing to my very large (400-500 students) introductory astronomy class, I generally try to wear a T-shirt that is relevant to the subject material. These include shirts showing (a) the curvature of space-time around a black hole, (b) telescopes at various observatories such as Keck, MMT, and HST, (c) the Milky Way Galaxy with an arrow labeled "You are here," (d) planets or the solar system, (e) famous scientists such as Galileo or Einstein, (f) specific types of astronomical objects (these can be obtained at conferences for specialists), (g) famous equations such as $E = mc^2$, (h) the Doppler Effect, (i) solar eclipses, etc. The possibilities are nearly limitless. You can even design your own T-shirts and order them through certain stores.

The students enjoy seeing these, and they like to guess what you will wear during the next lecture. If a student is having a hard time deciding whether or not to attend a given lecture, this little gimmick might provide the extra incentive that tips the balance in your favor. Upon seeing you, the student might be inclined to remain and listen attentively in order to find out how your shirt relates to the subject of the lecture. Finally, the atmosphere tends to be more relaxed if you wear T-shirts than formal attire, and the students feel at ease with the lecturer.

Alex Filippenko
University of California, Berkeley

Providing Some Incentives and Rare Opportunities

I give astronomy bumper stickers (like "Pulsars Turn Me On --- Fast," "Black Holes are Out of Sight," and "Red Giants Aren't So Hot") to students achieving the highest scores on my exams. These are available through *Sky and Telescope* magazine, planetaria, and elsewhere. Small meteorites, which can be purchased at Meteor Crater in Arizona and elsewhere, are also good rewards. Similarly, one can give out astronomy glow-in-the-dark ceiling stick-ons, posters, postcards, etc. (available at many observatories).

I also take some students to the observatory with me when I conduct research. Students can write an optional one page essay in which they describe why they want to go. My teaching assistants and I then read these to determine who will be the lucky winners. We look for some combination of burning desire, clarity, originality, motivation, etc. All expenses are paid for the students.

Alex Filippenko
University of California, Berkeley

The Seven Astronomical Data

I begin my astronomy class by asking students to generate their own list of celestial objects they have seen with their own eyes. Students soon realize that they already have some basic knowledge of the subjects that are part of the curriculum of this strange new course called "astronomy."

When students' lists are combined, the result is usually a brief but also fairly complete description of the observational facts available to a naked-eye astronomer. It often looks like this:

> Sky (blue in daytime and black at night)
> Sun
> Moon
> Stars (points with different brightnesses and colors)
> Planets ("wandering" stars)
> "Shooting stars"
> Fuzzy objects--comets or, perhaps, a faint nebula

I impress upon them that, at the very least, any astronomical theory developed later in the class must at all times predict these simple observations, if we are to follow the "rules" of science. While asking students what they see when they look up into the sky, answers such as "the Sun," "clouds," and even "birds" come up. These answers lead naturally into questions about the difference between day and night and the difference between astronomy and meteorology. They also anticipate a very important question in astronomy, namely, "How can we decide the distances to objects in the sky?"

Thomas Hockey
University of Northern Iowa

Troubleshooting Student Confusion

Professor Henry Gurr of the University of South Carolina describes a technique he has found very effective in troubleshooting student confusion. It is described in detail in *Physics Today* (August 1992) and may be very applicable to helping students during office hours.

The ways in which a student's thinking can go astray are legion and astonishingly unpredictable. When the student is confused, the instructor may feel a great urge to take over and show the student how to do the problem. But this urge must be suppressed."

"When a student...asks me a question, or when I find a weak conceptualization in the student's work, I put the pencil in the student's hand and say, 'Show me what you have been trying to do.' I conduct the session so that the student does all of the writing, calculating, and drawing, and nearly all of the talking. In this way, the information necessary to figure out what the student knows and doesn't know flows from the student to me. I can never get ahead of the student and cause additional confusion. When the student is doing all of the work of drawing, calculating, and explaining the problem, he or she controls the work speed so that a firm understanding is achieved at each step. The student will stop or falter when a point of confusion is reached, thus pinpointing the problem."

Greg Burley
University of British Columbia

Disturbing the Universe

One of the books I recommend to students who are interested in gaining some insight into the personalities of great scientists is Freeman Dyson's *Disturbing the Universe.* It provides excellent portraits of Feynmann, Teller, and other movers and shakers of modern physics. The book relates how scientists grapple with their societal responsibilities and discusses Dyson's role in WWII when he worked for the British Bomber Command. It also provides a penetrating view of how individuals in large organizations behave. This is a truly superb book for students.

Stephen M. Pompea
Pompea & Associates

A Testing Suggestion

Multiple choice is probably not the ideal form for tests; perceptive students often read into test items ideas that were not intended by the instructor. This sometimes leads students to conclude that two answers are equally valid or that no answer is correct. To minimize this problem, pass out a blank sheet of paper along with the test and tell students that when they find an ambiguous test item, use the paper to explain the ambiguity. Any student whose explanation convinces you that the writer understands the material in question will be given credit.

If you require that each explanation begin with a statement of which choice the student marked on the answer sheet, you will find that in most cases the correct answer has been chosen and you need not read further. Tell the students that their explanation must convince you that they know the material in question; you will not be overwhelmed with essays from each student. This technique has resulted in my students having a better attitude toward my tests and fewer "complaints" after them.

Karl Kuhn
Eastern Kentucky University

First Day Quiz

Over the years, in several introductory physics courses, I have given a very short (5 minute) quiz to get a discussion going. No one is really expecting a quiz on the first day, but in fact it is ungraded, and designed to point out some common strategies, makes it acceptable to students. Some discussions are very interesting!

The Quiz - It consists of three questions:

1. Within the space shown *draw* a triangle:

2. *Without* using a calculator, quickly find the SUM of the first 100 integers. That is, compute (write out the following for them)
1 + 2 + 3 + ... + 49 + 50 + 51 + 52 + ... + 98 + 99 + 100 = ?

3. The Atlas Cable Company must do the following: "On the south side of Tarrymore Road, the property owner wants us to hang a #32A-7 cable between the tops of two of our standard 10-meter long treated southern-pine poles. The length of cable between the tops of the poles is to be 18 meters, but the owner wants the middle of the cable to *just* touch the ground between the poles. Remember to sink each pole 1 meter into the ground for anchoring. Question: How FAR APART must the poles be to fulfill the order?

Here are some brief comments for each of the three questions.

1. No correct answers, but an illustration of *preconceived* notions. How many drew right triangles or an isosceles triangle? What about a general triangle (unequal sides, unequal angles)? How is the concept of a "triangle" stored in students' minds?

2. What strategies are useful? Looking for *patterns* is one of the purposes of this question. The picture is suggestive of grouping first and last (1 and 100, 2 and 99, etc.) which shows there to be 50 pairs, each with a value of 101, so the sum is 5050. However, number crunching isn't the idea here; a discussion about patterns can lead into ideas about order, simplicity, even "beauty."

3. Long question! But it serves to illustrate cutting through the extraneous *and* using common sense. Subtracting 1 meter from each pole leaves 9-meter-tall poles but with 18 meters of cable, and means we must plant the two poles next to each other if cable is to touch the ground as specified. This can serve as a stepping-stone for discussing ways to attack problem solving.

Robert G. Carson
Rollins College

Resource Books for Teaching

Two of the two best books of demonstrations for astronomy and physics classes are Jearl Walker' *The Flying Circus of Physics* (John Wiley & Sons) and Robert Ehrlich's *Turning the World Inside Out and 174 Other Simple Physics Demonstrations* (Princeton University Press). One of the best books on college teaching is the very practical book *Teaching Tips* by Wilbert J. McKeachie. It is subtitled "A Guidebook for the Beginning College Teacher".

Stephen M. Pompea
Pompea & Associates

Some Misconceptions about Astronomy and Astronomers

Exploring some these common misconceptions about astronomy might be useful ways to begin each week. Here are a few of my favorites misconceptions along with a sketchy explanation of the relevant facts.

Misconception: Astronomers spend all of their time looking through telescopes and recording their results.
Fact: Most data is taken electronically and in most telescopes there are no eyepieces. TV cameras hooked to guiding cameras are used for telescope pointing and control. The astronomer watches the TV monitor from the control room and takes data using an array of computers. Many astronomers work on producing computer models and rarely go to telescopes. The vast majority of an astronomer's research- related time is in reducing and interpreting data and not in observing.

Misconception: For faint objects, long photographic exposures are best.
Fact: Photographic film has largely been replaced by electronic detectors. Not only are the detectors more sensitive, they also allow the data to be taken directly into a digital format, where they can be processed by computers.

Misconception: Astronomers are old men in white coats.
Fact: Astronomers are a diverse group of people. Discuss with your class the role of women in astronomy and the recent meeting in Baltimore that gave rise to the Baltimore Charter.

Misconception: Space is empty.
Fact: Even though many areas of interstellar space have fewer atoms per cubic centimeter than some of the best vacuum produced on Earth, there is still a large variety of molecules in it, particularly in the molecular clouds. For example, large amounts of ethyl alcohol have been detected near the center of our galaxy. One large molecular cloud probably has more alcohol in it than all alcohol ever produced by humans.

Misconception: All stars are white in color.
Fact: Our color vision is very poor at night, so seeing the distribution of energy from a star is problematic. Photographic and electronic detectors reveal a wide variety of star colors or concentrations of energy in certain spectral regions. For some stars, the amount of energy at a number of wavelengths is similar; these stars tend to be viewed as white. Through a telescope where the stars are brighter and our eye's color-sensitive cones can be used, many stars have obvious colors of blue, green, yellow, orange, and red.

Misconception: Our Milky Way Galaxy is the only one, or at least one of the largest.
Fact: Our galaxy is rather average in size, being much larger than the dwarf spheroidals and significantly smaller than giant ellipticals. There are many thousands of clusters of galaxies that have been studied in some detail so our galaxy can hardly be called "unique".

Misconception: As long as city lights are far away, they don't affect telescopic observations.
Fact: Light pollution is a serious problem and affects many major observatories. Its effect may be apparent a hundred miles or more from a large city.

Misconception: We can see stars only at night.
Fact: Our sun is a star and is a major object of study for astronomers. Many features can be observed on the sun in detail that cannot be studied very well in more distant stars.

Misconception: The sun rises in the east and sets in the west.
Fact: Only on the first days of autumn and fall (autumnal and vernal equinoxes) does the sun rise due east and set due west. In the northern hemisphere summer, it rises north of east and sets north of west. In the winter, it rises south of east and sets south of west.

Misconception: The sun is overhead every day at noon.
Fact: For any spot on Earth north of 23.5 degrees north latitude or south of 23.5 degrees south latitude, the sun never gets overhead. If you subtract 23.5 from your latitude, this will tell you the how far from the zenith the sun is at local noon on the first day of summer. On this day the sun is about as high in the sky as it ever gets.

Misconception: Any high mountain is a good place for a telescope.
Fact: A high mountain usually ensures clearer air and less water vapor to interfere with infrared observations, but the steadiness of the air is also a prime consideration. Many observatories benefit from being in a desert region or where steady air movement from the ocean stabilizes the atmosphere.

Misconception: Passing through a comet's tail could destroy the Earth.
Fact: The tail of a comet has been described as "the closest thing to nothing something can be and still be something". The nucleus is a giant dirty snowball, but the tail is

very tenuous. A collision with a nucleus could be notable, but passing through the tail of a comet is not usually a newsworthy event.

Misconception: There is no gravity in space.
Fact: There is gravity in space just as on the earth. Where there is mass there is gravity. If there was no gravity, a satellite would not stay in orbit around the Earth, or for that matter would not fall towards the Earth into the Earth's atmosphere.

Misconception: A black hole will suck you in because of its incredible gravity.
Fact: A black hole may be formed from the collapse of a massive star. Its gravitational force at a distance is no larger than that of the original object. A black hole has the same "amount" of gravity as a star of that mass. If the sun were replaced by a black hole of the same mass, we would not notice any significant change in the earth's orbit, only in the amount of light.

Misconception: Astrology is based on science.
Fact: There is no scientific basis for astrology. It is a pseudoscience. The origin of sun signs and the attribution of characteristics due to a person's sun sign have no scientific basis.

Misconception: Constellations are meaningful.
Fact: Every culture has its own set of constellations. The creation of constellations is something anyone can do, with enough imagination. The establishment of constellations and constellations boundaries is largely an arbitrary process.

Misconception: The stars we see in the night sky are typical stars.
Fact: They are quite different in type and luminosity from the average star. They tend to be much more luminous than average. You would only recognize a few stars in a list of the 20 nearest stars, which is probably a more representative population of stars.

Misconception: The higher the magnification, the better the telescope.
Fact: Light-gathering power or collecting area is the primary function of a telescope. The image is blurred by the atmosphere and excessively magnifying it only gives a larger blurred image. The use of adaptive optics in telescopes will help recover the resolution inherently possible.

Misconception: Shooting stars are stars that have fallen from the sky.
Fact: The glow is given when small pea-sized chunks of material, which are in orbit around the sun, enter the earth's atmosphere and are heated

Misconception: There are billions of stars in our solar system.
Fact: Just one star, the sun, is in our solar system.

Misconception: If you are a Leo, the sun was in the constellation of Leo on the day you were born.

Fact: If you are a Leo, chances are that the sun was not in Leo when you were born. (It might have been in Leo 2,000 years ago on the day you were born, but many branches of astrology have not caught up to this fact.) The change is due to precession, the wobble of the Earth on its axis of rotation, a phenomena that has been known for thousands of years.

Misconception: The gravity from the planets exerts a profound effect at your moment of birth.
Fact: The gravitational effect of the doctor and nurse are far more significant. Perhaps we should form signs based on which side they were standing on at your moment of birth.

Misconception: The sun is incredibly dense.
Fact: The average density of the sun is 1.4 grams per cubic centimeter, not much more than water and less than most rocks. The center of the sun is quite dense.

Misconception: Constellations should look like the object or figure they are named after.
Fact: The constellations are named to honor great heroes or figures. Does the state of Washington look like George Washington?

Misconception: Stars must be round because that is how they look in a telescope.
Fact: The circular pattern, or Airy pattern (after astronomer George Airy) is a result of the diffraction of light waves by the round objective lens or mirror of the telescope. If the mirror was square, for example, a different-shaped star image would be seen in a high-quality telescope. The disk of a star is too small to be seen directly.

Misconception: All stars are the same size.
Fact: The largest stars (such as Betelgeuse) are thousands of times larger in diameter than the smallest ones, the white dwarfs.

Misconception: A black hole is black, truly invisible, and cannot be detected.
Fact: The collisions among the material going into a black hole (like water going down a drain) would create intense X-rays which could be observed. This accretion disk would be a signal for a possible black hole even if the object itself couldn't be seen.

Misconception: The Little Dipper and the Seven Sisters are the same.
Fact: The Little Dipper is a northern circumpolar group of stars and is visible year round. It is not very bright and is hard to spot. The Pleiades or Seven Sisters is a tight little group of stars that are part of the constellation of Taurus. It is visible in the winter and spring sky and is easy to spot when it is up.

Misconception: The North Star (Polaris) is the brightest star in the sky.
Fact: Polaris has a magnitude of 1.99. The brightest star in the sky is Sirius, which is at a magnitude of -1.45. Since each magnitude represents a brightness change of a factor of 2.5, Sirius is nearly 25 times brighter.

Misconception: The Big Dipper is a constellation.
Fact: It is only part of the constellation of Ursa Major, the Big Bear. Familiar groupings of stars like this are called asterisms. Another asterism is the Pleiades.

Misconception: Meteors can only be seen at certain times of the year.
Fact: Although there are a dozen or so major showers each year, sporadic meteors are visible every clear night and usually several per hour can be seen if the sky is clear and dark.

Misconception: The best way to see a faint object in a telescope is to look right at it.
Fact: Looking slightly to the side of a faint object will let you see it better. Looking straight at a faint object puts the image on a part of the retina which is optimized for seeing sharply and in color. Your best night vision is on another part of the retina. Glancing to the side of it will make it appear brighter.

Misconception: There is no need for more telescopes on the ground since space telescopes have made them obsolete.
Fact: The great expense of lifting a telescope into space and making it work in space makes space telescopes most useful for observing wavelengths of light that are absorbed by the Earth's atmosphere.

For maximum light-gathering power, 8 and 10 meter telescopes are now being constructed on the ground. The Hubble Space Telescope mirror is less than 3 meters in diameter. Ground-based telescopes are critically needed for their tremendous light gathering power. Adaptive optics has made it possible to make observations of objects that are nearly unaffected by the Earth's atmosphere. Ground-based telescopes with adaptive optics can create images that have the sharpness known only to space-based telescopes a few years ago. For a given aperture size, space telescopes are much more expensive and difficult to maintain than ground-based telescopes.

Stephen M. Pompea
Pompea & Associates

Astronomy and Music

These musical pieces in the table below have been used to introduce lectures for "Introduction to Astronomy" at the University of New Hampshire. The music is played during the last few minutes of the walk-in period prior to the beginning of class. The lecture usually starts with an explanation of the relationship of the music to the subject to be discussed. The music allows connections between different fields of knowledge.

Topic	Title of piece	CD/collection	Composer
Introductory lecture	Sunrise	Also sprach Zarathustra	Richard Strauss
Measurement, scientific method	Eyes of a Child	To our Children's Children's Children	Moody Blues
Stars in the sky	Nights in White Satin Spring: Aquarius Fall: Virgo	Days of Future Passed Zodiacal Sumphony	Moody Blues Richard Clayderman
Seasons etc.	The Seasons Spring Brahms	Meditation- Classical Vol. 7 Nepal Folk Songs	Tschaikovsky
Moon, phases etc.	Full Moon Hatha - Sun and Moon	From the full moon story Music for Yoga Meditation	Kitaro Tony Scott
Planet motion	Mercury	The Planets	Holst
Copernicus, Tycho, etc.	Venus	The Planets	Holst
Kepler's Laws	Mars	The Planets	Holst
Newton's Laws	Of Science	Also sprach Zarathustra	Richard Strauss
Gravitation	Heavenly Illusion	From the Full Moon Story	Kitaro
Uses of Gravitation:	Neptune	The Planets	Holst
Tides	Full Moon The Tide Rushes In	From the Full Moon Story This is the Moody Blues	Kitaro Moody Blues
Light, telescopes	New Lights	From the Full Moon Story	Kitaro
Light, colors	World of Rainbows	Canyon Trilogy	Carlos Nakai
Space Astronomy	New Horizons	The Seventh Sojourn	Moody Blues
Magnetosphere	Aurora Aurora	From the Full Moon Story From Heart to Crown	Kitaro Rob Whitesides-Woo
Planetary magnetic fields	Jupiter	The Planets	Holst
Atmospheres, greenhouse	Venus	The Planets Winds of Mars	Holst Bach/Mars Pathfinder
Water, life on other worlds	Remember when there was water	Carry the Gift	Carlos Nakai & William Eaton
Comets	Another satellite	Skylarking	XTC
Star magnitudes, etc.	Star Chant	Earth Spirit	Carlos Nakai
Black body radiation	Black is Black		Los Bravos
HR Diagram, Mass-Luminosity relation	The Sun is still shining	To our Children's Children's Children	Moody Blues

Topic	Title of piece	CD/collection	Composer
Energy source of the Sun (stars)	The Sun Is a Giant Ball of Incandescent Gas		There Might Be Giants Provided by student of the class and his band
Solar atmosphere, activity	Magic fields Corona	Meditation Sampler 2 Solar Trilogy	
Solar wind, interstellar gas	Wind Dance	Desert Dance	Carlos Nakai
Interstellar dust, star formation	Sunset Sunset	Days of Future Passed India	Moody Blues Kitaro
Aging of stars	Star Chant	Earth Spirit	Carlos Nakai
White Dwarfs, Novae	Sirius	Beyond Dreaming	Robert Haig Coxon
Supernova Pulsars, SNR's	any of CD Elements Cancer	Implosions Cycles Zodiacal Symphony	Stephan Mikus Carlos Nakai Richard Clayderman
Special Relativity	"Thinking is" The Best Way to Travel Future/Past	 Cycles	Moody Blues Carlos Nakai
Effects of General Relativity	Eclipse	Faces of the Harp (Narada Collection)	Andi Williams
Black Holes	Cleft in the sky	Canyon Trilogy	Carlos Nakai
Milky Way Galaxy	Mist	India	Kitaro
Galaxies	Spiral Passage	Canyon Trilogy	Carlos Nakai
Expansion of the universe	Explosions Polka	Ein Straussfest	Johann Strauss
Quasars, active galaxies	Thunder & Lightning Polka	Ein Straussfest	Johann Strauss
Cosmology I	Future/Past	Cycles	Carlos Nakai
Cosmology II, Early Universe	Creation Chant	Canyon Trilogy	Carlos Nakai

For more information on music to teach astronomy by see:
http://www-ssg.sr.unh.edu/406/music.html

Eberhard Möbius
University of New Hampshire

Chapter 2
Science and Pseudoscience

Scientific Creationism

Teaching in the Bible Belt, I get many questions about scientific creationism. Whatever your feelings about religion, these questions must be handled diplomatically or you risk losing nearly the entire class. These questions most commonly pop up during discussions of the big bang or the origin of the solar system. The person who asked the question is likely to already be firmly convinced that the science is incorrect, but if you deal with the question well you may convince the person who is sitting on the fence trying to deal with the same question. A few thoughts on dealing with such questions:

Be diplomatic. Never disparage anyone's religious beliefs even if you disagree with them. You will only alienate class members. Students are not going to give up lifelong religious beliefs because of anything they learn in a science class. You must therefore convince them that they can keep their religious beliefs as they learn science. Science does not address basic religious questions: Is there a God? If there is, why did he put us here? etc. Therefore, modern science is not incompatible with the world's major religions. Many Christian theologians prefer the Big Bang theory over the steady state because the Big Bang has a moment of creation. (Scientists of course have other reasons.) This position is in fact the official position of the Roman Catholic Church. The Genesis story is taken as metaphor by these theologians.

Scientific creationism as it is normally defined takes the Bible as being literally true, the final authority, and not subject to disproof. Scientific research is then conducted merely to find evidence supporting the literal Genesis story. This of course is *not* science, as all proper scientific ideas are subject to disproof.

Understanding the Universe by Phillip Flower (West Publishing) contains an excellent essay on this topic on p. 692 of the first edition.

Paul Heckert
Western Carolina University

Creationism in Genesis

It is a mistake for instructors to assume that only fundamentalist students from the boondocks will have creationist tendencies. The idea that the Bible contains a viable account of origins in Genesis is very widespread, held by tens of millions of Americans. Many students feel that this account should be presented alongside the standard big bang cosmology as an alternative model. Rather than just dismissing the biblical record as "mythological", "religious", or "poetic" as many texts are likely to do (and thus confirm to the students that they are unwilling to consider it), it is far more helpful to actually use the Genesis account to illustrate a specific pre-scientific (non-scientific) creation story. Most of the students are not familiar with the finer details of the account and will be surprised to discover the many pre-scientific ideas it contains.

These ideas include: (1) calling the moon a "light" with a status similar to that of the sun; (2) referring to day and night, and evening and morning, before the origin of the sun; (3) making the sun and moon "rulers" of the day and night; (4) stating that the heavenly bodies are for "signs" and "seasons" thus demonstrating the absence of a distinction between astronomy and astrology; (5) the placement of a solid sky in between the waters of the earth and the heavenly storehouse of waters. This sky is called "firmament," "dome," or "solid vault" in various translations. The original Hebrew work is *reqia,* which means "pounded out thinness," or "that which can be formed into a bowl." The inverted bowl cosmos is an excellent precursor to the discussion of Aristotelian crystalline spheres which, while somewhat more sophisticated than the Genesis account, still contains many mythological elements.

These ideas, and others, will facilitate the acceptance of contemporary scientific ideas by more conservative students. It is pedagogically efficient to displace some preconceptions to make room for the desired concepts.

Karl Giberson
Eastern Nazarene College

Earliest Science

This idea concerns the earliest evidence that science was being conducted by humans, and it can be introduced when the topic of seasons is discussed. The evidence goes back about four thousand years to the British Isles in the form of megaliths used for estimating the length of the year and predicting the summer solstice. Many sites exist where upright stones were used to sight the sun on the horizon returning to a particular position in the course of the year. This position was not the most extreme north or south position corresponding to the solstices because the daily change in the position of the sun can be much smaller than 0.1 for many days around the solstices. Changes this great correspond to much less than 20% of the apparent angular diameter of the sun, and were beyond the capability of early societies to measure. For this reason Stonehenge was not used as a center to determine the summer solstice; but was used to acknowledge it, after its time of occurrence was predicted by making observations elsewhere.

The position on the horizon measured with precision at many sites was in the general eastern, or western, direction, where daily changes in position are nearly the full diameter of the sun. But this direction does not correspond to the time of the equinoxes when the sun's declination is 0.0, but to the megalithic equinox where the declination is 0.7. The megalithic equinox corresponds to the midyear position rather than the mid-angular position. By determining the length of time to return to the megalithic equinox over a period of several years the length of the tropical year could be estimated with considerable precision. And by adding one quarter of the year to the time of the megalithic equinox, the summer solstice could have easily been one or two days off from the time we now consider as the summer solstice, it did not matter to the early societies, as long as everyone agreed to acknowledge the solstice at the same time.

Louis Winkler
Pennsylvania State University

Observation and Experiment

Strictly speaking, astronomy is an observational science: not all the variables are controlled (as in experiment) and the time at which measurements are made must also be recorded (since many interesting astronomical phenomena are those which do vary in time, even if over times longer than a single human lifetime).

Thus, though astronomy is classified as a physical science alongside physics and chemistry, since they tend more to deal in repeatable experiments astronomy is not exactly like them. In many ways it is more like meteorology (you had better record the time you measured temperature and wind direction) and geology, using what is seen now, in conjunction with laboratory data, to construe what went on in the past, and to predict what will happen in the future.

Anthony J. Weitenbeck
University of Wisconsin

The Shape of the Earth

Given our understanding of the nature of gravity, one of the best clues that the Earth is round is that we measure it to be flat. The only two cases in which this might be so are on an Earth that is flat but infinite in extent, or on an Earth that is roughly spherical.

Let us assume that the Earth were really flat, and of finite extent, such that one could "fall off the edge". What would we see? Well, we know that the gravity vector would always point to the true center of mass of the earth, and so as we walked further from the center we would have the sensation that the land was sloping more and more upward. As we got closer to the edge we would notice that the air was getting quite thin, and the slope more steep. At some point we would have to give up the attempt to reach the edge due to the lack of air and inhospitable climate.

On our way back home, we could follow one of the streams that flow down to a radially flowing river, to the vast ocean that occupied the entire central portion of our planet. On the shore of this ocean we could look out to the ships that disappeared over the horizon to trade with folks on the ocean's other side. We would conclude that we lived in a vast bowl, whose bottom was filled with water, and who knows what gods or demons live outside the bowl.

Of course on the other side, another society entirely could exist, separated from ours by unscaleable peaks, but having much the same world view. If one could leave the planet, and see the "real" situation he would discover that the planet was really a disk, with two mounds of water, one on each side. The surface of each ocean and of the air of the planet would follow the gravitational equipotential, in this case approximately a prolate spheroid.

How could an astronomer on such a planet discern the truth? Well, one possibility would be analogous to the measurement of the mass of Mount Everest on Earth. Comparison of the local gravitational vertical with the local astronomical vertical shows a component of the gravitational vector attributable to the mass of the mountain pulling a plumb bob out of alignment. Many other "gedanken" experiments can be devised to describe the world view on such a planet. For example, what would oceanographers measure in an ocean depth survey?

William G. Weller
Association of Universities for Research in Astronomy

Induction and Deduction -- Scientific Method

Although the method of science is most often taught by example (and hopefully by demonstration), it is useful to explain to students the logical systems that lie at the heart of the scientific method. Contrast deductive arguments as being truth-preserving and yielding certain conclusions (as in arithmetic), and inductive arguments as being probabilistic and not yielding certain conclusions. Point out that induction is the primary tool of the scientific method because it can lead to new knowledge. Arguments based on induction must be bolstered by a good statistical sample. Discuss how reliable the inductive conclusion that "All people are right-handed" would be based on examining samples of 5, 10 or 100 people.

It is also good to address directly the confusion between causation and correlation (which afflicts the scientific literature at times!). Bertrand Russell told a good story about a chicken that got fed every morning when the Sun came up, and learned to inductively associate sunrise with getting fed. The surprise came when one morning the Sun rose, and the farmer came out to wring its neck and put it on the dinner table. Finally, point out the majestic level of deduction involved in Newton's law of gravitation.

Chris Impey
University of Arizona

The Laws of Physics

Imagine a foreigner watching a game of American football. He knows nothing about the game, and he wants to learn about it by watching. For a while there is nothing but confusion for him; but later he will begin to see patterns that repeat, indications that there is organization, and evidence that there really are rules for the game. Of course, his task would be simplified if he had a rule book, or if someone would explain the game to him.

Astronomers want to learn about the game of reality in the Universe. They have no rulebook nor anyone to explain things to them. They must observe Nature and try to find repeating patterns, indications of organization, and evidence that there really are rules for this game. By a long process of trial and error, astronomers (and scientists in general) have written a rulebook for Nature, but this rulebook is tentative and subject to change and modification. The rules that Nature seems to obey are known as the Laws of Physics.

The Laws of Physics are not qualitative statements about what Nature does, more or less. They are mathematically precise statements that say what Nature must do. It is not at all obvious that Nature must be mathematical and obey these laws, but it has been found that, at least to a very good approximation and in a wide variety of circumstances, it does. For example, it does not rain when conditions are about right. If conditions are within the range specified by the appropriate laws, it must rain; otherwise, it cannot rain.

Suppose astronomers want to study stars. After gathering the required data, they build a model of a star. They do this by calculating the properties a star must have according to the rulebook on stars. Finally they compare the calculated properties with observed properties of real stars. If the comparison is poor (and if they cannot blame the observations!), then they see how they can modify the rules to improve the comparison. When the comparison is good, the astronomers gain confidence that their ideas of the Laws of Physics pertaining to stars are in good agreement with those that Nature uses to make real stars.

Thomas Swihart
University of Arizona

Scientific Method and Scale Errors

Despite a vast observational effort over the past 20 years, astronomers still deduce distances to nearby galaxies that differ by a factor of two. The reasons for this variation indicate how important it is to understand the errors attached to all astronomical data. One type of error is a scale error. Use of the more distant standard candles is tethered to the distances of nearby standard candles. For example, the distance to the Hyades is measured by a variety of methods, and the method of main sequence fitting then gives the relative distances to hundreds of open clusters. Any change in the distance estimate of the Hyades will therefore propagate outward and change the distance estimates for galaxies. In general, relative distances are better determined than absolute distances. Imagine you have a ruler, and have been happily measuring the distances between many objects in your house. Then the ruler manufacturer informs you that the markings on your ruler are smaller than true centimeters. All your estimated distances are suddenly too large, because of a scale error in your ruler. Note that you don't have to remeasure everything; you just have to reduce all your measurements by the ratio of the old and new centimeters. Similarly, any recalibration in a fundamental local distance implies changes in all larger distances.

Chris Impey
University of Arizona

Scientific Method-Random Errors

Distance estimates are also unavoidably affected by random errors. In the late 18th century, German mathematician and astronomer Karl Gauss demonstrated that characteristic uncertainties are associated with *any* measurement. It is a powerful result of statistics that as the number of independent measurements is increased, the precision of the result increases as the square of the number of measurements. For example, parallax can be used to measure the distance to a single nearby star, with a certain error. However, if the parallax can be measured to 100 stars in a cluster that are at the same distance, then the distance measure is 10 times more accurate. Similarly, observations of many Cepheids in a nearby galaxy like the Large Magellanic Clouds can be used to define a period-luminosity relation with greater accuracy, leading to a better distance estimate.

Using a ruler analogy, the accuracy of a single measurement is limited to a millimeter, the smallest division marked on the ruler. However, with repeated measurements, it is possible to make a measurement with accuracy better than a millimeter. A related issue is random sampling. Some distance indicators use the brightest star or planetary nebula as a standard candle. But stars and planetary nebulae have a range in luminosity, so how many do you have to observe before you can be confident that you have found the brightest? By the time an entire galaxy of stars is being considered, it is safe to assume that the rare, brightest stars are represented.

Chris Impey
University of Arizona

Scientific Method-Systematic Errors

Systematic errors are errors that do not diminish in importance with repeated measurements. Unlike scale errors, systematic errors need not affect all distances measured beyond a certain scale. Systematic errors usually affect the use of certain distance indicators or measurements made in certain parts of the sky. Historically, the biggest mistakes in determining distances have been caused by neglecting dust obscuration. The diffuse interstellar medium contains dust grains which dim the light from distant stars, making them appear farther away than they really are. The dust obscuration and the systematic error in distance increase with distance through the disk of the Milky Way. Since spiral galaxies are dustier than elliptical galaxies, systematic errors can also arise when using standard candles (Cepheids, supernovae) to compare the distances of both galaxy types. It is important to be confident of the basic physics of a standard candle. As mentioned before, Shapley was misled because he did not realize that Cepheids in Milky Way globular clusters are intrinsically less massive and luminous than Cepheids in the Magellanic Clouds.

As an analogy for systematic errors, imagine a metal ruler that expands when heated (made of copper, for example). Distances measured in your hot kitchen will be systematically larger than distances measured in your cold bathroom. Since the physical properties of the ruler are different in the kitchen and the bathroom, no amount of repeated measurements will remove this error.

Chris Impey
University of Arizona

Powers of Ten

Students often have trouble using the powers of ten or scientific notation. They wonder if the powers of ten were invented so teachers could have more good test questions. Students need to know that they were invented out of necessity. This is a dignified way of saying that they were invented out of laziness. Yes, scientists are lazy. We got tired of writing the distance to the nearest star as 6,000,000,000,000 miles, so we abbreviated that number by using scientific notation: 6×10^{12} miles. (We got even lazier and started using light-years, but that's another story.) Scientific notation is even more important for impoverished graduate students who don't have a lot of money to spend on paper and pencils.

This exercise is a lot of fun and illustrates well the extent of the universe in space and time as well as giving good practice in scientific notation. Since it is a bit like a card game it is a lot of fun. Here's how it works: Take some index cards and put numbers from -5 to +15 on them. Each student is dealt two cards. One card is the "time" card; the other is the "distance" card. The number on the card is the exponent for a number in scientific notation. For example, if a student draws 5 for the time card, the student must bring to the next class a picture of an event lasting 10^5 seconds. If he or she has a "distance" card that reads -2, then that person must bring a picture of something that is of order 10^{-2} meters in length.

At the next class, the students hang their pictures in order from shortest to longest time, and from smallest to largest sizes. Of course, the people at the smaller and larger ends of the scale will have a harder time than those in the middle of the range. At the upper end, 10^{15} meters is of order of 1/50th the distance to the nearest star, so this exercise is not dependent on students knowing very much about the size of the galaxy, distances to galaxies, and so on. The exercise can be repeated later in the semester with larger, or even smaller numbers.

Another variation, if you have a creative class, or a small class, is for them to demonstrate the events and distances in one representation or another. Obviously, the longer events will mostly be stills, but this can be fun for the class to guess the timescale represented. The exercise is a lot of fun for students and fosters an awareness that the human time and distance scale is quite limited.

You can modify this exercise to work with small children as well.

Stephen M. Pompea
Pompea & Associates

Exponential Notation

To demonstrate the difference between linear and exponential growth, and set the scene for the use of scientific notation, take two examples. First, challenge a student in class to fold a piece of paper 9 times over on itself (the student will manage 6 or 7 if he/she is strong). Second, tell the story of a boy who asked for a month of pocket money where he gets a cent on the first of the month, 2 cents on the second of the month, etc. By the end of the month, he has 2^{30} cents, or $10,737,418.24. Also, you can contrast normal (linear) and cancerous (exponential) cell growth in the human body.

Chris Impey
University of Arizona

Left-Handers

Take a head count in class; find out how many are left-handers (it will be 10-15%). Ask why so few. Guide the discussion that follows to show the critical thinking skills needed in science. Point out that in 1900, the OED counted only 2% left handers. Clearly we have a cultural bias, but where did it come from? Point out that apes do not discriminate, and that flints and cave painting show that early man was equally left- and right-handed. Note that the early Greeks favor the left (aristera = aristocracy), that Hebrew and Arabic are written from left to right, and that around 400 B.C., the change came, where the Romans gave us "sinister" and began the taint of left-handedness. Argue that it was sun worship among northern hemisphere civilizations that led to the dominance of right over left, since the sun appears to rise on the right as seen from the north (examples of auger in Rome, Buddhist prayer wheels, Mecca, right-hand references in Bible). Note that in the few southern hemisphere civilizations, there are good examples of left-handedness dominating or being equal (Hottentots, Samoans). Of course, the argument is somewhat tongue-in-cheek, but is meant to get the class thinking. Given these origins of the bias, why has right-handedness continued to dominate? Note that in primitive and warlike societies, left-handedness is an advantage, since it is correlated with good spatial awareness skills. For example, 90% of Olympic fencing champions are left-handed. The answer is that the cultural bias is perpetuated by right-handers, whose verbal skills are better. Left-handers have been brainwashed (finish with tongue firmly in cheek)!

Chris Impey
University of Arizona

Scientific Method

Use astronomy from Copernicus to Newton as a paradigm for the scientific method. Contrast the roles of Brahe as an observationalist whose data was crucial even though he didn't believe the Sun to be at the center of the solar system, Kepler who modeled the data of Brahe with ellipses although he didn't have a theory for such a model, and Newton as a theorist who synthesized a vast amount of data with a concept that had enormous predictive power. To counter the notion that pure science has no practical use, point out that Newtonian mechanics led directly to the Industrial Revolution in England, and hence to our wealthy society.

Chris Impey
University of Arizona

On the Nature of Science

After several years of teaching in the physics classroom it becomes all too clear that the majority of students have an incorrect view of the nature of science. Many have come to believe in science not on a basis of proofs and observations, but on the basis of faith. It's both fun and enlightening to use this aspect of their "scientific" training to befuddle and confuse them and make them take a second look at the nature of science.

Almost without exception students believe that the earth is spinning. When asked how they know, it usually comes down to a statement of faith in somebody or something: "My teacher said so" or "That's what my books said." Sometimes an enterprising youth will say "We have pictures from space." The appropriate response is, "Haven't you ever seen fake pictures of UFOs?" or "Do you really believe in everything that comes out of Hollywood? Do you believe in King Kong, the Mummy, etc.?"

Proving that the earth is spinning, is, for the vast majority, an impossibility. Take advantage of this ignorance of the scientific process to "prove" the contrary, that the earth does not spin. The ultimate goal is to show that we cannot depend upon mere reasoning or faith in anybody to know whether or not something is a fact or not. Use the classical Aristotelian proofs of the non-motion of the earth in the following manner:

If the earth is spinning, why aren't projectiles left behind? If I stand still and throw an eraser up in the air, it comes back down. If I throw it up into the air and I'm moving (throw eraser up and then begin walking), the eraser lands behind me. Clearly, when I stand upon the earth and toss the eraser up it falls straight back down to me. The earth, therefore, is not moving.

If the earth is spinning, why don't I feel the motion? When I ride in a car I feel the motion; when I ride a bike I feel the motion. If the earth is spinning at 1,100 mph at the equator and 700 mph at my latitude, why don't I feel the motion?

If the earth is spinning, why don't I feel a prevailing wind blowing continually from the east? When I walk briskly or ride a bike or drive my car, I can feel the wind blowing in my face. If the world is turning (at 700 mph in my latitude), why don't I feel a wind blowing continuously from the east?

If the earth is spinning, why aren't the oceans cast off from the equator into space? If I ride a bicycle through the pouring rain, the water flies off at the circumference of the wheel (the fastest moving part, just like the earth's equator). The earth is spinning much faster at the equator than is the tire. Earth still has its oceans and, therefore, it must not be spinning.

It's nice to end lectures with this subject matter. Any physics student will see the fallacy of these "proofs," but the vast majority of the students will be left in a near state of shock! Use this shock as a motivational tool for studying Newton's laws of motion. Once these laws are examined and explained, return to these "proofs" and pick each one apart.

Carl J. Wenning
Illinois State University

Listening to Water Boil

At the beginning of physics classes, we gloriously tell students that science is founded on observation, and that science begins with a careful description of a phenomena, with a suspension of our preconceptions. How good are we at applying this? How many people have really heard the sound of water boiling, or more correctly the different stages in the boiling of water. If you think that water doesn't make any sound until it starts to boil, then you haven't really listened.

If you have a class too large to gather 4 or 5 deep around a watched pot you'll need a microphone. Have the students listen carefully as water is heated from room temperature to rolling boil on a hot plate. Have the students record their impressions and threaten to make them turn them in at the end of class. This will help them to take the exercise more seriously, which will make the room quiet enough to do the experiment. You may have to boil several times, so everyone can hear the distinct stages that are present. I won't describe the sounds in each stage because I want people to approach the observation without an "answer key."

This exercise is very useful in the early part of an course, when we are trying to convey the perceptual root of all of the sciences. It also can serve as a good introduction to the serendipitous aspects of science since many things in a field have been discovered by youngsters (newcomers) who had minimal preconceptions about the structure of knowledge.

Stephen M. Pompea
Pompea & Associates

Science and Pseudoscience

After examples of scientific method in action, stress that theory can never be proven. Of 99% of data explained, 1% can force a new theory (perihelion of Mercury). Public attitudes towards pseudoscience can be summed up by the story of Bohr. After he won his Nobel prize, he showed a New York Times reporter around his impressive lab in Copenhagen. As they went in, the reporter was surprised to see a horseshoe over the main entrance. "As a great physical scientist," the reporter said, "surely you of all people are not superstitious." "I've heard," said Bohr, "that it works whether you believe in it or not." Hone critical thinking skills with examples from UFOlogy, von Danikenism, astrology (note confirmation bias in psychology), and creationism.

Chris Impey
University of Arizona

Confronting Pseudoscience

Confronting pseudosciences such as astrology can be made a part of an astronomy course. Student interest in this area is certainly high. The instructor can best make his/her points about astrology by not lecturing in an authoritarian manner, but by using an experiment (which is supposed to be how science works). One "experiment" which I have used successfully in confronting astrology is to have the students not consult a horoscope for three days. During this period, students keep a daily diary of their feelings and moods. They also make note of how their financial and love lives are going, since this is a major aspect of horoscopic predictions. In class I distribute the 36 horoscopes of the three-day period, but with the sun signs removed and the order scrambled. Students must now find the three that describe their days best. Then we compare their answers with their "correct" horoscopes, and see if the predictive power of astrology is significant. This is also a good time to talk about precession, the Age of Aquarius, and how the people who are sexy Scorpios may not really be Scorpios after all.

Stephen M. Pompea
Pompea & Associates

Scientific Method-Round Earth

Challenge student's knowledge that the Earth is round. Ruling out passed-on knowledge or evidence such as photos from Apollo 8, how would students experience the shape of the planet? If someone offers the example of ship's masts disappearing over the horizon, counter with the known bending of light in mirages. If a lunar eclipse is offered, point out that a single observation does not distinguish between a disk and a sphere. Note that for much of history, the limited distances traveled by most people stopped them noticing the curvature of the Earth (later on, use this analogy to argue that the curvature of the universe is not observable unless you travel large distances also).

Chris Impey
University of Arizona

Chapter 3 Observational and Historical Astronomy

A Demonstration of Parallax Measurements

To perform this demonstration, it is required to find a fairly long unobstructed area with a view of the horizon, or at least a very far-off building. A football field or campus mall are good locations. Materials required are a measuring tape, and a large protractor, such as those used for blackboard geometry layouts. A tripod or table on which to place the protractor is helpful, but not necessary.

First, lay out a baseline of 10-meters length perpendicular to the line of sight to the horizon, and mark each end. This represents the diameter of the orbit of the earth. Have a student walk some distance away, say around 100 yards, and remain standing for the duration of the following measurement. He will represent the star to be measured. One could also use a tall stick painted white for this.

Set up the protractor at one end of the baseline, and measure the angle between the student (star) and some prominent feature on the horizon, such as a tree, power pole, or building. Repeat the measurement from the other end of the baseline, using the same distant reference object. A little geometry will show that the angle subtended by the baseline as seen from the student (star) is just the sum of the two angles thus measured, if the reference object is at infinity (i.e., very far compared with the star.) Use this fact to calculate the distance to the student, then verify with the tape. An interesting discussion here is to determine the effect of having the reference and measurement star at similar distances, as is sometimes the case in astronomy.

Sending the student much farther off will reduce the angle, increase the errors and demonstrate the difficulty of making actual stellar parallax measurements. There is also the possibility of having all students make the same measurement, and using statistical analysis to derive the "best" estimate of the distance. By inverting the measurement, that is measuring the baseline from the star, one can define a "pardeg" or parallax degree (that is the distance at which the baseline subtends one degree) to demonstrate how the parsec is defined.

William G. Weller
Association of Universities for Research in Astronomy

A Sky Briefing

Once a week, I have a student or students present a "sky briefing" for the rest of the class. This five-minute presentation, modeled on a weather forecast, "predicts" astronomical events that can be seen in the sky during the upcoming week. The phase of the Moon, planetary alignments, and meteor showers are usually included, as well as rarer events that happen to occur that week (e.g., eclipses). Occasionally, late-breaking news such as a nova or the appearance of a new comet can be added. The knowledge that class members are to prepare a briefing gives relevancy to my discussion of astronomical tools such as sky charts and almanacs. Accessing various astronomy information "nets," in order to prepare their briefings, gives students an opportunity to practice their computer literacy skills, as well.

This exercise serves two purposes. First, it emphasizes the dynamic yet orderly changes that are observable in the heavens, even to the unaided eye. Both the changing and cyclical nature of the night sky are things that many students have not thought about. Second, the briefing allows students to work on vital public-speaking skills in a practical setting (as opposed to an exercise in an oral communications class) and do so in a relatively low anxiety environment.

Thomas Hockey
University of Northern Iowa

Projection Effects in Extended Radio Source Structures

Several extended radio sources (jets, lobes) show C- or S-shaped symmetries. These structures may only be projection effects, i.e., a source may be a spiral, or a slightly curved C-shape, but if aligned in a certain way, we see it in projection as an S or sharply curved C. In both cases, the appearance of only one jet or lobe may be due to projection effects as well.

An excellent way to illustrate this concept is to take a length of wire and bend it around to pattern the actual shape of the jet (either a curve or a spiral) and then move it around so as to give the class different orientation views. This allows those students who have difficulty visualizing things in three dimensions to see what's going on. (Original idea by Ann Gower, used with permission.)

Joanne Rosvick
University of Victoria

Counting the Visible Stars in the Sky

This exercise can be done on a dark, moonless night at the observatory. Have the students roll an 8"x11" sheet of paper into a cone. The diameter of the cone times 4, divided by the length of the cone, gives the fraction of the sky that can be seen through the cone by looking into the small end. A student may see about 50 stars. By doing a count in various directions, the student can then average the number of stars seen. The student then multiplies the average by one over the fractional part of the sky being viewed by the cone, and this gives the approximate number of stars (3000 - 6000) that can be seen with the unaided eye.

V. Gordon Lind
Utah State University

Measuring the Size of the Sun

After discussing the distance to the sun and its angular size in the sky, students can be taught how the diameter of the sun can be calculated. At this point the students should be asked to measure the angular size of the sun by using a pinhole camera. The "camera" can be made by making a small hole in a piece of paper such as with a paper punch. The students can do this out of the classroom before the next lecture period. They measure the size of the sun's image and the distance from the "camera" to the sheet of paper onto which the image is cast. The ratio of those two numbers is the same as the ratio of the sun's diameter to its distance, so in effect, the students measure the sun's linear diameter. As the students bring their results to class, the various results can be discussed and eventually averaged to give a fairly reasonable answer for the actual size of the sun.

V. Gordon Lind
Utah State University

Music of the Spheres

A powerful idea in the history of astronomy is the concept of the "Music of the Spheres." Point out that the fundamental link between mathematics, music and astronomy was first put forward by Pythagoras, none of whose writings survive in original form. Pythagoras believed that the essence of the universe was mathematics, and that harmonies of the whole numbers governed the motions in the sky (just as the ratios of the whole numbers 1,2,3,4 govern music, i.e. 2:1 octave, 3:2 fifth, 4:3 fourth). Note that this idea is strikingly modern, since cosmologists once again believe that the entire universe is governed by mathematical laws. Pythagoras developed geometry and the idea of the five "perfect" solids, that is there are only 5 three-dimensional figures whose faces are polygons. Two thousand years later, Kepler used this Pythagorean philosophy wholesale, trying to fit the orbits of the planets with nested Pythagorean solids (and succeeding to better than 5%!), and writing extensively about the "Music of the Spheres" Rodgers and Ruff have actually recorded a realization of the "Music of the Spheres," see the article in the May/June 1994 issue of *American Scientist* (the recording can be purchased from the authors).

Chris Impey
University of Arizona

Subdividing Time

Many issues of astronomy are covered with a study of time and how humans have chosen to divide it up over the ages. The fundamental problem of calendars is of course the fact that the lunar (synodic) month of 29.53059 days does not divide evenly into the solar (tropical) years of 365.242199 days. Human culture has divided on this issue, with desert-dwelling Arab cultures choosing to follow a strictly lunar calendar where the holidays migrate through the year, and the great agricultural civilizations following a solar calendar. The division of the lunar cycle into weeks has been followed by many cultures; our version follows the Romans' in a 7-day cycle named according to the seven visible celestial objects in the sky (Sun, Moon, Mars, Mercury, Venus, Jupiter, Saturn -- the order set by ancient astrological tradition).

Get students to be aware that the strict astronomical cycle is broken in our Germanic-influenced culture. Tuesday, Wednesday, Thursday, and Friday are named after the Norse/Teutonic gods Tiw, Woden, Thor, and Frigg. The division of 12 hours for day and nights has its origin 4000 years ago in the bright timekeeping stars, or decans, that the Egyptians used to subdivide the night sky. The units of minutes and hours stem from the Babylonian number system, and of course are chosen to accord with the degree unit of angular measurement, since Babylonian astronomer-priests were the first to keep careful records of stellar and planetary motions. Finally, it can be pointed out that the only rational and decimal calendar in human history was the French Revolutionary calendar, with 12 equal 30-day months, five festival days and a leap year. Each day was divided into 10 hours, each with 100 minutes. The calendar lasted only 12 years before being nullified by Napoleon with the Pope's blessing.

Chris Impey
University of Arizona

Right Ascension and Declination

There are two exercises I use near the beginning of the semester to help students understand the celestial coordinate system with Right Ascension (RA) and Declination (DECL). Both of these exercises use the SC1 Constellation Chart, published by Sky Publishing Corporation (49 Bay State Road, Cambridge, MA 02238).

The first of these exercises helps the student learn to use RA and DECL. The coordinates of a number of objects are given to the student, and the student is asked to locate them on the chart and write the names of the objects. Then the names of some other objects are given and the student is asked to determine their RA and DECL. These coordinates can be checked by referring to tables, such as those included in most introductory astronomy textbooks. When the students are finished with the exercise, a few minutes are spent describing the various objects on the chart.

The second of these exercises is a plot of the positions of a planet throughout the year. The coordinates for a planet (I have used Venus) are taken from *The American Ephemeris and Nautical Almanac* issued by the Nautical Almanac Office, United States Naval Observatory, and available from the US. Government Printing Office. I have used data for 1975 for Venus, but other years and other planets could be used. By observing the path of the planet on the chart, it is clear that the planet is always near the ecliptic. Retrograde motion is apparent.

David Blewett
Butler County Community College

Sun's North-South Movement

To observe the sun's north-south movement with changing seasons, ask the students to measure the length of the shadow of a prominent pole or building at 1-week intervals for several weeks. The sun's motion becomes obvious. The measurements should be made at the same time of day each time.

Similarly, the students can be instructed to observe the position on the horizon of the rising or setting sun.

V. Gordon Lind
Utah State University

Star Motions

Imagine going into a room and standing under a light that is suspended from the ceiling. As you spin round and round, the objects in the room appear to move, appearing and disappearing, in the same way the stars do in the sky. If you were to look overhead as you turn, the lamp over your head would appear to stand still and things near it circle round and round. Now, the North Star, Polaris, stands over the earth's "head." Stars near the North Star circle round and round, never disappearing from view. These are the circumpolar stars. Other stars rise and set. These equatorial stars are too far from the pole star to remain visible all the time.

Carl J. Wenning
Illinois State University

Measuring the Size of the Earth

This is a fun exercise which duplicates one of the most "classic" science experiments of all time: Eratosthenes' measurement of the size of the Earth. Mr. E measured the earth by noticing that the sun shone directly in a well at noon at Syene, Egypt on the first day of summer, but was not overhead at noon at a more northerly location. The method really only requires that the altitude of the sun be measured at the same (relative) time (preferably local noon, when the sun is highest that day) from two different locations on the earth, on the same day. One also needs to know the distance between two points. The greater the separation between the two points on the earth, the easier the experiment is. Mr. E used two points that were 5,000 stadia apart (the Greek unit if length was the *stadium*, which is equal to a little more than 600 feet, depending upon whose stadium you used. This unit is still used in surveying, and is defined as so many *chains*).

The teacher needs to find another group, somewhere, that would like to do the same experiment. Most astronomy teachers know colleagues in other parts of the country who would be glad to cooperate. The class also needs an accurate way to measure the altitude of the sun. A crude, but instructive way is to use a giant protractor or to use trigonometry and use the shadow of a vertical pole. Of course, a sextant could be used. If the two locations are several hundred miles apart, then the accuracy of the measurement can be somewhat less. This is a good time to discuss measurement errors and ways to minimize inaccuracies and maximize accuracy and precision. The class could divide into two groups and one group could drive 50 or a hundred miles. If this is done, it is advisable for the groups to be due north and south of each other so the measurements could be made exactly at the same time. (It is easier to synchronize watches than to determine when the sun crosses the meridian.)

This experiment takes a bit of preparation, but it has a tremendous payoff for students in the realization that the ancient Greeks were pretty smart after all. Which reminds me, it was only last year in Arizona where a state official proclaimed that teachers should teach the flat earth theory if some of the school district's parents believed in it, in order to give equal time alternative views of "sensitive" issues.

Stephen M. Pompea
Pompea & Associates

Four Minutes' Difference Between Sidereal and Solar Day

After illustrating that a year consists of 1 more sidereal than solar day, then it is easy to show that 24 hours (1 day) divided by 360 equals 4 minutes/day: (24 X60) / 360 = 4

But even more pedagogical and interesting to the students is for them to observe a star or constellation rise 28 minutes earlier a week later, or 1 hour earlier in two weeks. Therefore, assign them to make this observation with some instructions on how to do it.

V. Gordon Lind
Utah State University

Astronomy and Historical Events

Dr. Donald Olson of Southwest Texas State University has written a number of articles relating history, literature, and astronomy that can be used very effectively with students. He discusses the astronomical aspects of important historical events in articles such as those listed below. They provide very dramatic introductions to lectures as most students are curious about these events. A few of his articles include:

D. W. Olson, "The Tide at Tarawara: November 20, 1942--A Historically Significant Apogean Neap Tide," *Sky and Telescope* **74** (No. 5), 526 (1987).

D. W. Olson and R. L. Doescher, "Van Gogh, Two Planets, and the Moon," *Sky and Telescope* **76** (No. 4), 406 (1988).

D. W. Olson and M. S. Olson, "William Blake and August's Fiery Meteors," *Sky and Telescope* **78** (No. 2), 192 (1989).

D. W. Olson and R. L. Doescher, "Lincoln and the Almanac Trial," *Sky and Telescope* **80** (No. 2) 184 (1990).

D. W. Olson, R. L. Doescher, and E. S. Laird, "Columbus, Regiomontanus, and the Sky of January 17, 1493," *Sky and Telescope* **81** (No. 1), 81 (1991).

D. W. Olson, "Pearl Harbor and the Waning Moon", *Sky and Telescope* **82** (No. 6), 651 (1991).

D. W. Olson and R. L. Doescher, "Paul Revere's Midnight Ride," *Sky and Telescope* **83** (No. 4), 437 (1992).

These articles can be used very effectively with an accurate computer planetarium program such as *Voyager*, *Redshift*, or *Starry Night*, to show the class the positions of celestial objects on particular days. An overhead projection screen for the computer is very effective for large group presentations.

Stephen M. Pompea
Pompea & Associates

Determining Local Latitude and Longitude

Knowing global latitude is one of the most important variables for ocean-going vessels. A sextant is a device that measures the altitude of almost any object, a building, a flagpole, a tree, etc. Using simple trigonometry, the height of an object can be determined by stepping backwards from the object until the sextant reads 45 degrees. Then the height of the object is the same as the distance away from the object plus the height of the observer's eyes. Alternatively, the height can be determined by finding the product of the tangent of 90 minus the measured angle and the distance from the object. Used at night, local latitude is the value as the altitude of the North Star, Polaris. /

Major ocean-going explorations needed to measure local longitude. In order to do this, navigators used very accurate clocks, or chronometers, to keep track of the local time in Greenwich, England. The sun moves west across the ocean about 1 degree every four minutes. By measuring how many hours, minutes, and seconds noon occurred on the ocean AFTER it occurred in Greenwich, navigators could determine exactly how far west they were located (their longitude).

Today, people use time zones to keep "clock" time close to solar time. Universal time (UT), Greenwich Mean Time (GMT), and Zulu Time (ZT) are terms used to describe the local mean time in Greenwich, England. People that live 7.5 degrees East and 7.5 degrees west of Greenwich, England all use the same "clock time" or "time zone." This 15 degree sliver is called the Zeroth Time Zone. The 15 degree sliver that ranges from 7.5 to 22.5 west longitude is the First Time Zone. The 15 degree sliver that surrounds 90 west longitude (from 82.5 to 97.5) is called the Central Time Zone.

As the Earth rotates, local noon moves westerly across the Earth's surface. It is first noon in New York, then St. Louis, then Kansas City, then Denver, the Los Angeles, and, later, Hawaii. Local Noon moves westerly at 4 minutes per each 1 degree. To be more accurate, this needs to be corrected using the *Equation of Time* in order to account for the variation in Earth's orbital speed around the Sun. Ancient mariners needed an accurate clock (chronograph) to keep track of Greenwich Mean Time so that, by noting the difference between Noon in Greenwich and their local noon (or sundial noon), they could determine their local longitude with high accuracy. It was in fact the need for accurate clocks that slowed worldwide ocean exploration until the Renaissance.

Directions:

1. Construction of the sextant. Hang a paper clip or washer or taped penny with a 20 cm string from the center hole of a large protractor. Tape a sighting "straw" to the straight side of the protractor and invert as shown in the diagram below.

2. Making measurements with the sextant. Sight the tops of trees, flagpoles, or stars through the straw; the string should hang down vertically. Hold the string against the protractor and read the value. The altitude is 90 minus this measured value .

3. Determining height with sextant. Method "A" is to walk away from the object until the sextant reads 45 degrees. The height is simply the distance from the object plus the height of the sextant. Method "B" is to sight the top of the object being measured from a known distance away. The height is simply the height of the sextant plus the product of the distance away and the tangent of the altitude.

4. Determining local latitude (conducted at night). Sight the north star, Polaris, through the straw. Again, the altitude is the measured value minus 90 degrees. Make the measurement three times and average the results. The altitude of the star Polaris is the same as your latitude. Check your results with a road map or GPS surveyor .

5. **Set a watch or clock to "exact" time.** On public radio, morning news programs start at exactly the top of the hour. CNN often shows the exact time during their broadcasts. Alternatively, short wave (HF) radio station WWV in Colorado broadcasts "official" time signals. The beginning of news radio broadcasts are probably the easiest way to set your clock to "exact" time by resetting the second timer as it starts.

6. Calibrate a sundial or mark a gnomen's shadow. Mark the top of a gnomen's shadow every 15 minutes or so from 10:30 am until 1:45 p.m.; the more often the better. Record the exact time each shadow top is marked.

7. Determine the clock time that a sundial shows as "local noon." Connect each of the points to make a curved, smooth arc. Local noon occurs when the shadow is the shortest; interpolate between marks to determine the exact time of local noon. Due to variations in the Earth's orbit around the Sun, the result is most accurate in mid April, mid June, early September, and late December.

8. Calculate local longitude. The Sun moves across the sky at a rate of 4 minutes per degree. Subsequently, if the local noon occurs 24 minutes after the top of the hour, then the observer is located (24/4=6) eight degrees west of the center of her time zone (or 96 west longitude).

9. Compare your results with a road map or GPS surveyor.

Tim Slater
Montana State University

Chapter 4
The Moon and Eclipses

Scaling the Earth-Moon System

The large appearance of the full moon in the sky often is deceiving to students who assume it is very large and relatively close to the earth. To dispel this misunderstanding, make a model of the moon from a Styrofoam ball that is approximately 1/4 the diameter of an earth globe (as the kind used in geography). Spray paint it gray and attach a string to it that is approximately 30 times as long as the diameter of the earth globe.

Attach the other end of the string to the earth globe and have a student walk until the string is fully stretched. The distance between the moon and the earth on this scale is surprising. When held side by side, the globes also reveal the tiny size of the moon relative to the earth.

James E. Kettler
Ohio University - Eastern Campus

Eclipse Measurements and the Earth-Moon Ratio

Students are given a photo of a partially eclipsed moon. They are informed that this was a full Moon and the portion not illuminated is in the Earth's "umbra." Using a compass, they are to complete the circular umbra, matching the curved portion across the face of the Moon as closely as possible. Students then measure the size of the Moon's diameter and the "umbra" diameter to ascertain the approximate ratio of the Earth and Moon in a manner similar to the ancient observers.

Appreciation for early astronomy is enhanced when you remind people that the first observers did NOT have photography; they had to make their own pictures to measure.

Charles Roberts
S.U.N.Y. at Morrisville

Line of Nodes

To illustrate the line of nodes and lunar eclipses, mount two circular disks (at an angle of 5° or exaggerated) and place a ball in the center (the earth). One disk is the earth-sun plane, the other is the Moon-Earth plane. With the earth-sun plane horizontal, walk around a sphere (the sun) and show the moon's position over a year and line of nodes points at the sun twice a year.

Neal P. Rowell
University of South Alabama

Simulating Solar Eclipses

This demonstration uses the cutouts shown in the diagrams above, on an overhead projector, to show partial, total, and annular solar eclipses.

The pieces are cut from poster board. Card A and Card B are 30 cm square, to fit the top of the overhead projector. A 12-cm circular hole is cut from the center of Card A; this will represent the sun.

Card B begins the same as Card A, but the hole is enlarged irregularly to represent the solar corona. The "large moon" is 12 cm in diameter; the "small moon" is 10 cm. Attach small loops of masking tape to the two moons to act as handles.

Put Card B on the projector and lay Card A directly on top of it; turn on the projector and the sun appears on the screen.

To show a partial eclipse, place one of the moons at the side of the hole in Card A and slowly slide it in a straight line across the hole, off-center so it never covers the hole completely. On the screen you will see the progress of a partial eclipse as the moon "takes a bit" out of the sun, covers it partially, and moves off. Various percentages of coverage can be demonstrated by covering more or less of the sun. To show a total eclipse, place the large moon at the edge of Card A and slide it in a straight line across the hole so it covers the sun completely and drops into it. At this point, remove Card A, leaving the moon and Card B in place; the corona is then visible on the screen, surrounding the black moon. After a few seconds of "totality," replace Card A and slide the moon away.

An annular eclipse is demonstrated by using the small moon, sliding it directly across the hole, but leaving Card A in place. It will not fill the hole completely, so a ring of light appears on the screen as in an annular eclipse.

To accurately simulate an eclipse, the moon should be moved in a straight line at a constant speed. Also, it is helpful to draw guidelines on Card A to follow as

Simulating Solar Eclipses

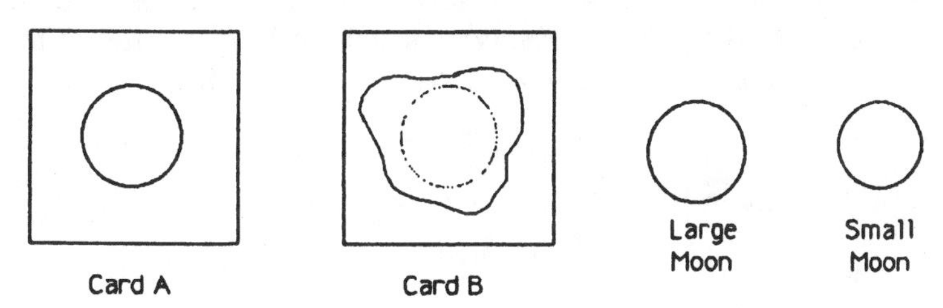

you slide the moon, especially for the total eclipse. Guidelines can also be drawn for various percentages of coverage in a partial eclipse. As an added touch, cut notches in the edge of the large moon to simulate the diamond ring effect.

Tom Damon
Pikes Peak Community College

Scaling the Earth-Moon System

One can scale spheres and distance of separation of the earth and moon. I used an ordinary earth globe from the geography class which had a diameter of 9 inches. On this scale, the moon diameter would be 0.27 times as small, or 2.5 inches within an error of + 3% of the value 2.43 inches required. A Styrofoam ball can easily be obtained of this size. I painted it dark gray to represent the moon.

The distance from the center of the earth to the moon is approximately sixty earth radii. So a string was cut to this length using the 4.5 inch radius of the earth globe above. The string would then be 22.5 ft. long. When the moon globe is held beside the earth, its really small size is apparent. When the moon globe is placed at the end of the string, the large distance from the earth is evident. One can mention the time taken to travel this distance by the astronauts and the velocities involved in the trip, etc.

James Kettler
Ohio University - Eastern Campus

Phases of the Moon

Styrofoam balls (about 4 inches in diameter or larger) are handed or thrown out to the students. These represent the Moon. An overhead, projector light bulb is used as the Sun. The students can move the balls around their heads to see the changing phases. The students can also stand up and rotate themselves to show the times of the rising and setting of the Moon in various phases. We have enough balls for about half the students, who must throw the balls to the other half after their turn. The point is made and students enjoy it.

John B. Laird
Bowling Green State University

Synchronous Rotation

Why is it that although the moon is orbiting the earth and also rotating that we earthlings always see the same face? The reason, of course, is "synchronous rotation." Synchronous rotation can be explained by a blackboard diagram. However, it is much more effective to use two students, one as the earth and one as the moon. There is a psychology involved here. Students may be drowsy and inattentive, but if you ask one or more of their peers to get out of their seats and come forward and take part in a demonstration, you immediately have 100% attention. It may be a slight "fear" that they will be next -- I'm not sure of the nature of the psychology, but for the next few minutes, everyone is watching closely.

The first student's part is to play "earth." He/she simply stands still. The moon-student faces the earth. That person's face can be the side facing the earth. If the "moon" now starts revolving and at the same time rotating, what combination will keep the same face inward? Best understanding occurs if the moon-student goes through a quarter rotation. Then another quarter revolution followed by a quarter rotation, etc. Of course, both motions happen together. In synchronous rotation, what is synchronized? The rotation is in synchronization with the revolution.

David B. Porter
Portland Community College

The Moon's Orbital Motion

After illustrating in class with Styrofoam balls or with drawings on the chalkboard of the moon's motion about the earth, ask students to confirm this motion by watching the position of the moon for several consecutive evenings at the same time each evening. The motion becomes obvious and often really startles the students.

V. Gordon Lind
Utah State University

To Show the Moon's Orbital Shape

To show the moon's orbital shape, photograph the moon at a 3-day intervals. After 1 month, develop the film and show the variation in the moon's diameter, which illustrates how the moon is closer and farther away at different times in its orbit. With a little help, the students can convert the changing diameters to changing distances, and thus, actually plot the orbit.

V. Gordon Lind
Utah State University

Chapter 5
Earth, Solar System Mechanics

A Model of the Earth's Magnetic Field

I use a variation of the old iron filings and magnet demonstration to help students visualize the Earth's magnetosphere. First, I place a sheet of clear plastic, with a picture of the Earth drawn on it, over the glass surface of an overhead projector. (The plastic is important added protection; I would never hear the end of it from my colleagues if I allowed iron filings to fall into the bulb or fan motor and shut down the overhead!) I then center a bar magnet on the drawing. This is followed by a layer of clear plastic wrap and then, finally, a sprinkling of iron filings. A slight shake of the stack causes the filings to line up with the magnetic field lines. If I choose the sizes of the drawing and magnet carefully, a roughly to-scale silhouette reminds my students that the real terrestrial magnetic field is not generated by giant bar magnet!

A final note: When the demonstration is over, I simply lift off the plastic wrap, thereby speeding up the retrieval of the iron filings.

Thomas Hockey
University of Northern Iowa

Why Planetary Orbits Are Nearly Circles

When discussing the nebular theory for the formation of the planets, it is helpful to have a tub of water which is stirred vigorously with a stick giving the water all kinds of random motion. A few seconds after removing the stick the random motion disappears (cancels out) and only a net circular motion remains. Thus the students can understand why the orbits of the planets are all in the same rotational sense and almost perfect circles.

V. Gordon Lind
Utah State University

Sun-Earth-Moon Motion

Early on in most astronomy courses, the instructor talks about revolutions and orbits, and the distinction between the two is one of many concepts that students have trouble with. A human demonstration makes these ideas clear and gives the student a sense of the scales involved. Ask for volunteers to act as the Sun, the Earth, and the Moon, and ask them to animate themselves in a model of the orbital motions of the three bodies (having introduced the material beforehand). It is most effective if you do not coach them, but let other students call out for changes (some will get so involved that they get up and forcibly move the people in the correct motions). Do not stop until you have them moving with more or less the correct relative orbital motions and spin rates. It's silly, but instructive, and it makes several important points in a way that dry words or slides or overheads cannot.

Chris Impey
University of Arizona

Planetary Orbit Analogy and Demonstration

Many students have difficulty understanding why gravitational force does not succeed in pulling planets into the Sun. The idea of elliptical orbits only adds to this confusion, for it is not clear to most people why a planet -- once it has moved closer to the Sun -- moves farther away again. Obtain a large bowl with a spherical surface (perhaps a salad bowl), and roll a marble around its inside surface to demonstrate the idea. Because of friction, of course, the marble will finally end up in the center, but I find that the students can be made comfortable with the idea of the lack of friction for a planet moving through space. The demonstration helps people relate ideas of inertia and centripetal force as well as illustrates that a certain specific speed is necessary for a given orbit.

Karl Kuhn
Eastern Kentucky University

Parallax via Two Eyes

One method we use to judge the distance to a nearby object is dependent upon our having two eyes. This can be demonstrated dramatically with no special equipment. Ask your students to pair off and have one student hold up a finger at eye level in front of the other person. The second person should then extend one finger downward well above the first person's finger and bring it down until the two fingers touch. This, of course, is easy to do.

Next, have each person close one eye as they repeat the procedure. They will probably be surprised that they have lost most of their depth perception and cannot easily touch fingers on the first try.

You can try this without the second person by using one finger from each hand, but although closing one eye causes a loss of depth perception, you know the length of your arms and it is easy to do. If you extend a pencil up from one hand instead of a finger, however, it fools you enough that touching the top of the pencil is no longer automatic.

The use of two eyes for depth perception depends partly on parallax, but another factor is involved -- each eye sees a slightly different view of the object itself. I'm not sure of the relative importance of the two factors; an experiment to test it should be fairly simple.

Karl Kuhn
Eastern Kentucky University

Precession Phenomenon Within the Solar System

The piece of equipment I use to demonstrate the precession phenomenon within the solar system is a heavy-rimmed gyroscope with two degrees of rotational freedom. Precession helps explain the changing pole star, drifting of the equinoxes, and the difference between zodiacal constellation and astrological signs. The device also helps explain the complexity of the lunar ephemeris, especially regarding eclipse patterns.

Louis Winkler
The Pennsylvania State University

Retrograde Motion of the Planets

Ptolemy's system for explaining retrograde motion of the planets is often difficult for students to grasp; I think part of the problem is that the system runs counter to their Copernican prejudices.

To make the Ptolemaic system easy for students to visualize, I use a frisbee with a large dot painted on it near the perimeter. I draw on the chalkboard the standard Ptolemaic system for, say, Mars, making sure that the epicycle for Mars is exactly the same size as my frisbee. I then place the frisbee over the epicycle. Now by rotating the frisbee while translating it along the deferent, I can show explicitly the motion of Mars.

A second suggestion involves the same frisbee. Aristotle argued from observations of lunar eclipses (among other things) that the Earth is spherical. To emphasize the point, I use a slide projector with no slide in it to shine a white light on a screen. (With newer projectors, you may have to insert an empty slide mount to defeat the automatic "dark slide" device.) A basketball or volleyball can be used to represent the Earth; its shadow on the screen clearly has a circular profile. Does this single observation of a circular shadow prove that the Earth is round? Certainly not: the frisbee is also capable of projecting a circular shadow. To distinguish by their shadows a frisbee from a ball, observation of different eclipses from different orientations of the Earth must be made: only the ball shows a circular shadow from all possible orientations.

The third suggestion comes from a teaching assistant, Susan Hubbard (to whom it should be credited). She was trying to get a student to visualize the idea that lines of sight from different points on the Earth to a distant object are approximately parallel if the object is far enough away. (The specific context was explaining Eratosthenes' method for measuring the circumference of the Earth, but the situation occurs quite frequently elsewhere in astronomy.) Walking to the far end of the room, Susan said, "Look at me. Are your eyeballs looking in the same direction?" "Yes," replied the student hesitantly. Susan then marched up nose-to-nose with the student. "Now look at me," she demanded. "Are your eyes crossed?" "Oh," said the student, "no further questions."

Philip Sakimoto
Whitman College

Seasonal Variation of the Solar Insolation

The familiar change in angle of altitude of the Sun with the seasons can be shown by using a standard earth globe that is mounted with a tilt of 23 $^1/_2$ degrees from the vertical. A flood lamp is placed on the center of a table and aimed horizontally. By placing the globe so that it is tilted toward the light, the analogy to the summer season is evident. Conversely, placing it so that the globe is tilted away from the light, the winter analogy is shown. Even the day and night variation in length is possible to show if the room is dark enough. A variation of this method is to place two thermometers in a pan of soil for each. If the one pan is illuminated by the flood lamp with slanted rays, the thermometer will not register much rise. The other pan is illuminated with more direct rays and the thermometer will register a higher temperature.

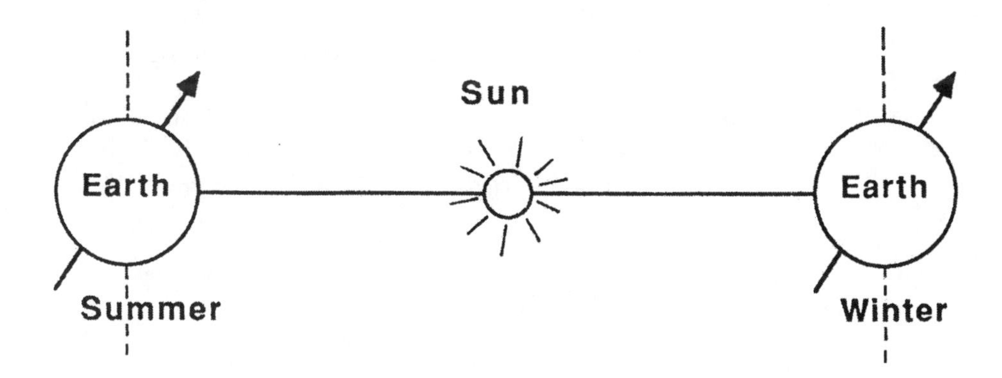

James Kettler
Ohio University - Eastern Campus

Bishop Ussher

The story of the Anglican Archbishop Ussher, confidant of King Charles I and internationally noted scholar of the 17th century, is a valuable one to relate to your class when discussing the conflict between dogma and scientific exploration. Ussher constructed a chronology of the Earth's history based on the Bible. His chronology was published in 1650 and 1654 and is notable in its specificity; even today some aspects of his chronology appear in some Bible margins. The bishop concluded at "Upon the entrance of the night preceding Sunday, October 23, 4004 B.C. heaven and earth were created." The Deluge followed 1,655 years later with Noah entering the Ark on Sunday, December 7, 2349 B.C. and emerging on Wednesday, May 6, of the following year.

Ussher's "scholarship" is notable not only because of its inaccuracy, but rather for its inability to adjust to new information. When European travel to China brought back information that the lineage of Chinese noble families had been traced back to times before the date of the creation of the world, no adjustment or serious questioning of Ussher's dates took place. Indeed, in many churches today these dates are given as gospel truth. This chronology was used later to dispute the scientific data that showed the earth to be much older.

Stephen M. Pompea
Pompea & Associates

Mars' Synodic Period

To see why Mars has the longest synodic period, it is best to imagine cars on a racetrack. It doesn't take long to lap the driver who is broken down alongside the road and one is passed rather quickly by the driver whose car is only slightly slower than your own. In this case, Mercury is the "supercharge" fast planet and Pluto is the "broken-down" slow planet. Mars is the planet moving only slightly slower than our "spaceship Earth"; thus it takes longest for us to pass each other in our race around the Sun.

Glenn Spiczak
Bartol Research Institute

Tidal Heating of the Moon and Planets

When the Moon was much closer to the Earth, the differential tidal forces on it heated it substantially, probably leading to volcanic activity. The concept of tidal heating can be illustrated very deftly using a soft rubber ball. Drill a small hole to the ball's center and bring along a thermometer and take the temperature of its middle (room temperature, hopefully). Then give the ball to some of the athletes in the class who will squeeze it until they hurt (5 minutes is OK). Then retake the temperature of the ball. It will have gone up considerably. Mechanical energy of macroscopic motion has become thermal energy expressed in microscopic motions.

Stephen M. Pompea
Pompea & Associates

Cratering Equilibrium

Demonstrate that a surface being cratered by projectiles can reach an equilibrium condition where its surface appearance does not appreciably change with further bombardment. In the lunar regolith such a steady state condition is believed to exist for small craters. If the number of craters created in one size class is ten times greater than the number created in the next larger size class, bombardment of the surface beyond a certain point will not change the nature of the surface. Therefore, information on surface age is lost beyond that point.

The demonstration can be done with projectiles of different sizes impacting on a sand surface. Experiment with different bombardment conditions (drop different size distributions of projectiles and drop some projectiles from different heights) to establish when a surface equilibrium condition may result.

Stephen M. Pompea
Pompea & Associates

Retrograde Motion

A demonstration for retrograde motion of the planets can be built. On a wooden stand, a series of gears (gear ratio about 3:1) is mounted. Turning a crank moves the gears by carrying two planets around a central point (the Sun) at different rates. A bar connecting the two planets indicates the line-of-sight from one planet to the other. A small arrow can attach to either end of the bar using a thumb screw. The students can see the retrograde motion by the direction the arrow points to.
Also, if a small penlight projector is attached to the arrow, the motion can be seen on the walls and ceiling of a darkened room. (This idea belongs originally to Morris Davis at the University of North Carolina.)

John B. Laird
Bowling Green State University

Cratering

It's amazing to most students that a small meteorite can create a crater many times larger than itself. Of course the falling object has tremendous kinetic energy which can be converted into cratering energy. Interestingly, a unit of energy from a TNT explosion will create the same sized crater as the same unit of falling kinetic energy from a fast moving meteorite.

The process of conversion of kinetic energy into a large crater can be illustrated nicely using mud or dry cement. Unless you have a ready source of mud handy, I'd recommend the cement. Place a tub of dry cement mix or equivalent on the ground. Then drop things (such as baseballs) on the cement mix. Amazingly, fairly large craters will be created. These craters will obey the same depth to width relationship that is found for lunar craters. Another amazing thing is that they will have central peaks, just like lunar craters. Experiment with different kinetic energies (velocities) and note the types of craters produced.

The craters that you make can be used to start a discussion of cratering histories. You can also experiment with the direction of the impacting body and how it affects crater morphology. Much can be learned about craters and cratering processes from playing in the mud. Students really enjoy the creative and destructive aspects of this demonstration.

Stephen M. Pompea
Pompea & Associates

Size of the Solar System

I use a football field analogy to show the size of the solar system:
Put the sun on the goal line, then

Mercury is on the 1-foot line
Venus is on the 2-foot line
Earth is on the 1-yard line
Mars is on the 1.5-yard line
Jupiter is on the 5-yard line
Saturn is on the 10-yard line
Uranus is on the 20-yard line
Neptune is on the 30-yard line
Pluto is on the 40-yard line (at maximum distance)

On this same scale, the nearest star would be some 500 miles away!

John Burns
Mt. San Antonio Community College District

Scale of the Solar System

When discussing Newton's Laws, I try to impress upon the students the range over which these laws operate. (Never mind cosmic distances, the solar system is vast enough.) The best way to help students to understand the scale is for them to walk it off themselves. In the back of astronomy books are the distances to the planets and the size of the sun. These numbers are only meaningful if they can be integrated into the students' experience. A good scale to use is one foot = one million miles. This makes the sun about one foot in diameter, close to the size of a basketball. It makes the planets small objects you can hold in your hand. It's a good idea to calculate with your students the sizes and distances of several of the planets, before letting them calculate the rest. You may find that this is a hard calculation for many students, until they see a few examples.

The earth, for example, would be 93 feet away from the basketball, easy to step off as paces. Ask for a volunteer for each planet and the sun, but be sure to pick a fast runner for Pluto; it's pretty far away. Have each student-planet hold an object equal in size to the planet at this scale or draw the planet to scale on a 3 X 5 card. Then they can walk out to the proper distance. Remember to put some planets on opposite sides of the sun. Once every planet is at the right distance, ask your students why it is hard to visit Jupiter or Mars. Be sure to mention that the nearest star, in the Alpha Centauri system, would be another basketball, far away. Have them calculate how far.

A good way to get a laugh is to tell your students that you want only one more thing before class is dismissed. You want each "planet" to make one revolution around the sun. There are lots of other points which can be illustrated, such as the fact that Pluto is so cold because of how far the sun is as viewed from Pluto. Ask students to imagine what the sun looks like from Pluto.

Stephen M. Pompea
Pompea & Associates

Motions of the Sun, Moon, and Earth

The idea I'd like to submit is a layered transparency device which I devised for demonstrating to students the motions of the sun, moon, and earth on an overhead. It consists of three overlaid sheets of transparency material appropriately sized, two of which rotate about a central point. The sun and the terrestrial orbit on the bottom sheet (in red) are fixed. The earth (in black with horizon of a specific point indicated) and the Moon (in green) on separate sheets can be rotated about the point in the center of the Earth.
This allows demonstration of the Sun and Moon rise or set and an understanding of how east and west are defined from the given horizon. Even though it is not to scale, it seems to help with these ideas.

Michael E. Mickelson
Denison University

Chapter 6
Newton and Gravity

Human Tides

Many pseudoscientific theories have claimed a cause-and-effect relationship between the moon and various earthly events. Astrology and biorhythms, for example, use cycles close in duration to the lunar period. As evidence for the claims these pseudoscientists often point to the earth's moon-generated tides. A quick "back-of-the-envelope" calculation can be done to find the tidal effect experienced by humans due to the moon. For maximum effect we assume the moon is overhead. The forces experienced by a bit of body mass m due to the moon of mass M at head and toe are:

$$F_{head} = GMm/r^2$$

$$F_{toe} = GMm/(r+h)^2$$

$$F_{head} / F_{toe} = (r+h)^2 / r^2$$

$$= 1 + 2h/r + h^2/r^2$$

$$= 1 + 10^{-8} + 10^{-17}$$

We would need 9 digits of accuracy to measure the effect. Put another way, the effect is that of one part in 100 million! Are human tides a large effect? Obviously not.

Eric Kincanon
Gonzaga University

Videostrobe Water Drop Gravimeter

Students can gain a fundamental experience of measuring the speed of falling bodies by watching television! Karen and Kenneth Brecher describe some aspects of using the stroboscopic properties of the television's sweep rate to measure falling objects or objects in periodic motion. The 60-Hz sweep rate of the television is used to freeze the motion. Full details are given in *The Physics Teacher*, February, 1990.

Stephen M. Pompea
Pompea & Associates

Galileo's Leaning Tower of Pisa Experiment

Will a feather fall as fast as a rock? Will a piece of paper fall as fast as a book? Most students will say no and when asked why not will comment on the air currents and the resistance on the feather or paper due to the air.

The instructor then drops a piece of paper and a book together, side by side, and thus easily verifies the students' expectations. The paper flutters around and takes much longer to fall. Then the instructor talks about what would happen in a vacuum. Most students will believe that the paper and book will fall together, although some students aren't sure. Without going into a vacuum, the instructor can strengthen the students' ideas by crumpling the paper and dropping it with the book side by side. They both fall pretty much together. The real clincher, however, is to place the flat piece of paper on top of the book and drop them together. The paper stays firmly in place on the top of the book all the way down. This makes real believers out of the students.

V. Gordon Lind
Utah State University

Simulation of Clustering and Accretion

There is an effective demonstration of some of the principles of gravitational attraction and accretion that can be done using an overhead projector. Get a wooden ring about 1 foot in diameter (the kind of hoop that is used for needlepoint is best because it has a clamp). Attach a piece of cling film moderately tightly across it. Uniformly sprinkle sand or small BBs over the surface of the film, and adjust the overhead projector so that the particles are in sharp focus. Then give the outer edge of the ring a series of gentle taps. The grains of sand or BBs will start to deviate from a random distribution as uneven numbers create small indentations which in turn attract more particles. Of course, the BBs will eventually end up as one mass in the center, but this can be avoided by inserting a support to raise the center of the film slightly. This will create a hollow of circular symmetry which will lead to several clumps along its circumference. The feedback of this mechanical setup is a good analog for the processes that concentrate matter from one region into a compact object (a star or galaxy for example), or that sweep up material over a wide range of orbits (as in planet formation).

Chris Impey
University of Arizona

Gravity and Orbits

Students often don't understand how a planetary orbit is produced. Specifically, if asked what would happen to Earth if the Sun's gravity were suddenly cut off, they respond that Earth would fly radially away from the Sun (rather than along the tangent to the orbit at the point of release).

Here's a great way to demonstrate what would actually happen. Get a bagel or a donut, and tie the end of a long string around one section of it. Hold the other end of the string, and carefully begin swinging the donut around over your head. This is analogous to a planet orbiting the Sun, with the string acting like gravity. Tell the students that they should closely watch the donut. Eventually the string will work its way through the donut, thereby setting it loose; students should note the direction in which it flies away.

To be effective, you need a long string; otherwise, the distinction between a tangential and radial path will be difficult for students to distinguish. However, with a long string the velocity rapidly becomes high, and the string quickly eats its way through the donut before the audience has had a chance to absorb the action. Using a partially hardened stale donut helps delay the release (but the student who catches the donut won't get a very tasty snack).

A bagel works quite well, but try not to use one so hard that it can hurt someone. With practice, you can learn how to adjust your pull on the string so as to launch the bagel into the audience, rather than towards the chalkboard.

This demonstration obviously brings a load of laughs, but I am convinced it is also instructive. Students who have witnessed it perform significantly better on the relevant section of the exam than those who have not. (I learned this trick from a graduate student who had been a teaching assistant for Robert Kirshner at Harvard. I don't know where Kirshner learned it; perhaps he conceived it himself.)

Alex Filippenko
University of California, Berkeley

Newton's Laws Illustrated with Stick and Balls

(Most physics classes have much more elaborate, quantitative demonstrations of Newton's laws of motion, but many astronomy students don't take physics courses and would probably choke on quantitative demonstrations anyway.)

Materials:

1.5-inch wooden dowel
3 balls, all the same size:
high density (e.g., metallic)
medium density (e.g., wooden)
low density (e.g., Ping-Pong)
1 balloon

First Law: Put a ball on the table. Note that it doesn't move. This is the first half of the law. Now roll the ball (use the high-density ball). Note that it moves at a constant speed across the table. (Show by pushing a book on the table it comes to test quickly because of friction. Observe that Newton, in deriving these laws for the Moon's orbit was dealing with a frictionless environment, and that's why he was successful.)
Second Law: Allow the rolling ball to roll off the table (catch it if you're quick enough) and point out the motion is with uniform velocity (speed plus direction) unless acted on by an unbalanced force -- the motion after it leaves the table.

Using the stick like a pool cue, poke one of the balls. Point out that the change in motion is in the direction of the applied force. Now set one of the balls moving , poke it with the cue at an angle to its motion and point out the change in motion is in the direction of the applied force, but the new motion is a combination of the former motion and the change.

Next, hit the dense ball with the dowel used as a baseball bat. Use a light whack and observe a small change in motion; use a heavy whack and observe a large change in motion. Note that the magnitude of change in motion (acceleration) is proportional to the magnitude of the whack (applied force).

Next, line up the three balls in a row in front of you, dense farthest from you. Hit the side of the table with the stick a few times. Point out these are at constant units of force defined as "1 whack." Now hit the dense ball with "1 whack" force; note its change in motion is small. Next, the medium-density ball; quickly try to catch it before it leaves the table. (By now, the students will be anticipating what happens next. Caution them on the dangers of Ping-Pong ball injuries, but reassure them that your department has sufficient liability insurance.) Hit the Ping-Pong ball. Point out that each ball had a different mass and that the change in motion is universally proportional to the mass. (Here you can write the $a = F/m$ equation).

Here is a good place to remind students of the difference between mass and weight. Point out that in a weightless environment, the relative masses of things can be ascertained by trying to shake them. Note that these shaking experiments were actually performed by Skylab astronauts.

Third Law: Hit the table with the stick -- hard. Tell them the sting you feel in your hand is a result of the third law reaction force. Blow up the balloon. Note that the elasticity of the balloon tries to compress the air, but when you let the balloon go, the air inside will be accelerated out of the back (where the hole is) and as a result of the third law, the balloon will accelerate toward the front. As you let the balloon go, it will fly on a wild path.

Point out that although the air is accelerated out the back, because the balloon doesn't have any stabilization vanes, where the back actually is keeps on changing. Note that this is the rocket principle and the breakthrough in rocket design was getting stabilization vanes on them.

John Broderick
Virginia Polytechnic Institute

Motions of Celestial Objects in Terms of Newton's Laws

Motions of celestial objects are understood in terms of Newton's laws. Many excellent demonstrations for Newton's laws are in physics lecture demonstration handbooks. A few of my favorites follow:

Use a lead brick and a hammer to demonstrate the concept of inertia. Put your hand under the brick and hit the brick with the hammer -- hard. The inertia of the lead will keep the lead from moving much and protect your hand. Then ask for volunteers to try it without the protection of the brick's inertia. One can also set the brick on an empty aluminum soda can or paper drink cup. With practice, one can hit the brick without smashing the can or cup. Then remove the brick and prove that the can or cup will indeed collapse from the hammer blow. Intellectual honesty then compels me to admit to the class that an additional contributing effect is the hammer blow being spread out by the brick, so the inertia is not the only effect.

To show Newton's second law, then hit the lead brick and drink can with equal force from the hammer. Point out that the less massive object had more acceleration. Park Plastics in Linden, New Jersey, makes a toy water rocket that is available at fine toy stores everywhere. It provides a fun way to demonstrate Newton's third law or the conservation of momentum. First fire it without water and act disappointed at the result. (After spending a whole week's allowance!) Then try it with water. Get yourself and the front row wet, but be careful not to hit anybody. Explain the dramatic difference either in terms of the third law or conservation of momentum. To show how gravity provides a centripetal force and results in circular motion, twirl a ball on a string. The string's tension is analogous to gravity.

Show with a plastic gallon milk jug that an object in free fall is weightless. Make a small hole in the bottom (a soldering iron works) and plug the hole with a cork. Fill the jug with water. Climb up onto a table, remove the cork, and drop the jug. Before letting go, point out that while the jug is falling, no water will leak out. The time is short. There are 3 landing options: 1) Allow the jug to splatter on the floor and people in the front row and to generally make a mess. The back rows will enjoy this option, but it will not ingratiate you with the custodians or the front row. 2) Hit the trash can. This is less dramatic and also less messy. 3) Try to hit the trash can but miss and hit the edge. This produces the biggest reaction from the class and also the biggest mess. You can, however, claim to the custodians that the mess was accidental.

Conservation of angular momentum is important in explaining the nebular theory for the formation of the solar system or the rapid rotation rate of pulsars. Illustrate this idea with a spinning stool. Have a volunteer sit on the stool with weights in his (her) outstretched hands. Start the student spinning, then have him pull his arms in. A figure skater starting a spin is also a good analogy.

Paul Heckert
Western Carolina University

Newton's Third Law

A more exciting demonstration is a rocket sled. A pressurized CO_2 canister (the size of a small fire extinguisher) is opened while sitting on a sled on wheels. The effect is quite dramatic and is also a lot of fun. A student can be chosen to stop the sled, or the "rocket" can be fired in the opposite direction. Donning a crash helmet for the demonstration adds a bit to the mood. One or two students (e.g., the one who volunteered to stop you) can be given a chance to try it themselves.

John B. Laird
Bowling Green State University

Angular Momentum

A stool which is free to spin and a couple of 1 or 2 kg weights are sufficient for this well-known but useful demonstration. I usually get a student volunteer to sit on the stool. I spin the student at a low speed, first with his/her arms held in near his/her chest (moving his/her arms out slows them down), and then with their arms out initially (and them moving them inward). In the latter case, one must be ready to steady or catch the student when they speed up. The student should move their arms in and out several times in both cases to make the point clear. I find this demonstration much more effective than the ice-skater illustration, since it's acted out in their presence, the change can be repeated several times (arms in and out), and the student speeding up in the second case can be quite startling and funny.

John B. Laird
Bowling Green State University

Tides

One of the most difficult ideas for the student is that tidal forces produce tides on both sides of the Earth. Some seem convinced from the argument shown in the following figure. The figure shows the Earth in orbit around the Sun (for example). At the first position (circle in lower right), three points are marked A, B, and C. If there were no central force, all would move in a straight line (indicated by the dashed line).

In fact, all are pulled inward, but A harder than B, and B harder than C. Since A is pulled harder, it moves inward more. C, being pulled the least, moves the least. The result after moving to the second position is an elongated shape. It is clear that both A and C must move away from B.

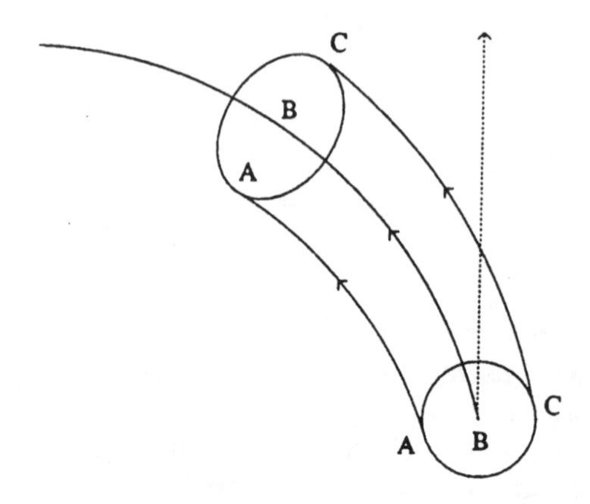

John B. Laird
Bowling Green State University

Center of Gravity

Although this demo is nothing new, the approach may be. It's a well-known fact among physicists that a person standing against a wall with his or her heels against the wall cannot bend over at the waist (knees straight) and pick up a piece of paper or other small object. The center of gravity will move out as the student bends forward causing him/her to rotate away from the wall. When introducing center of gravity I begin by asking who would like to win a $100 bill and pull one from my packet (or the largest bill I have) and place it on the lecture desk. I then proceed with the lecture and let the class think about the $100 bill on the desk. After introducing center of gravity I then let volunteers try to win the money. I haven't lost yet.

Bob Bechtel
Vincennes University

Newton's Laws and the Egg Drop

This is an attention-getting demonstration of Newton's Laws in action. It consists of a plastic container of water, topped in turn by an aluminum pie plate, toilet paper roll, and a raw egg at the very top! Place this set-up at the edge of the demo table so that the pie plate hangs over the table edge. Then, with foot on bristles of broom, tension the broomstick away from the set-up and let go of the broom handle! The handle hits the pie plate hard -- but the egg doesn't fly out into the audience (to their amazement and relief), but drops harmlessly into the water.

Constance Walker
University of Arizona

Newton's Cars

This is a great demo for Newton's 3rd Law (and a great lead-in to F = ma). Distribute a couple dozen 3"x 1/2" wood dowels, one 2 1/2"x 3 1/2" wood block, one 1"x 1" wood block; string, rubber bands, 3 nails and one match. The wood dowels lay the foundation for a "track" upon which the car will travel. Place 1 nail in the middle of one (2 1/2") end of the bigger block. Place other 2 nails at other end of bigger block near opposite corners; place elastic band around those 2 nails. Draw the back band toward the 3rd nail (it should not quite reach).

Tie a short piece of string around band and 3rd nail so string is taught; place smaller block on inner side of elastic band; place dowels like train rails in a single row with dowels 1" apart; place blocks on top; place lit match against string! The string will burn through, the rubber bands will send the small block flying one way and the car will roll the opposite way down the "track".

Constance Walker
University of Arizona

Chapter 7 Planets, Their Satellites, and Comets

The "Face" on Mars

Over the past few years, questions about the "face" on Mars seem to have replaced questions about Von Daniken and Velikovsky. To deal with these questions and to convince students that such formations can occur naturally, I point out naturally occurring geological formations on Earth that in some way resemble a human being. A local formation with which the class is likely to be familiar works best.

Teaching in western North Carolina I use the example of Grandfather Mountain, which my students tell me is named for its resemblance to an old man. (I'll admit I haven't seen it.) I also once saw a formation in southwestern New Mexico called the Kneeling Nun. From a distance it resembles a nun kneeling before an altar. (So far I haven't yet had a student counter that this formation proves the aliens must be Catholic.) If there are no good local formations use the example of patterns in clouds, which can resemble just about anything.

Paul Heckert
Western Carolina University

Relative Composition and Density of the Planets

Students have a difficult time comprehending relative densities without having a chance to experience them firsthand. Therefore, it is useful to prepare some objects of similar size and shape such as block cubes or spheres, but of different densities. For example, a steel ball of the same size as a tennis ball or a Ping-Pong ball really gets students thinking when they pick them up and toss them a little bit. To make balls of exactly the same density as the planets would be ideal but not easily achieved, but balls of steel, marble, wood, air, water (a plastic ball full of water) etc. are easy to come by.

Using the sense of touch and feel is a nice experience for the students and this exercise is very instructive in helping them to appreciate the differences among the planets. After the densities of the planets are experienced firsthand, the students are more responsive to discussing the various compositions that give the planets their densities.

V. Gordon Lind
Utah State University

Relative Brightness Intensity of the Sun to the Earth

To help the student appreciate the relative brightness of the sun to the earth a demonstration can be done in the classroom. Use a light bulb and a BB. Place the BB 110 times the diameter of the light bulb away (about 1 meter). Turn out the lights in the room and let the students observe the brightness of the illuminated face of the BB compared to the brightness of the lamp itself. The contrast is amazing and is certainly remembered.

V. Gordon Lind
Utah State University

Creating a Comet

Students often have a difficult time understanding how a comet can have such a low albedo but have such a high volatile content. I find that a demonstration in which we build a comet in the classroom works well. This is harder to do in Hawaii, so we make small comets by buying a cooler full of shaved ice, but in cold climates, a large cometary masterpiece can be created with everyone helping. The most interesting aspect is to take the white snow and begin slowly mixing the dirt with the snow -- to see that indeed a small mass fraction of dirt dramatically reduces the albedo.

I add a bit of the dry ice I pass around the room, so that the comet "sublimates." If this is done early in the lecture, and the room is warm (and if you didn't add too much dry ice which keeps the comet too cold), the melted snow will have dramatically darkened the surface of the comet, and the students can see quite readily how loss of volatiles near perihelion passage can create a very dark comet. This also leads to a nice discussion of what happens to the dust that does leave zodiacal light, and what will eventually happen to a comet if it turns off (cometary aging, confusion with asteroids, splitting, meteor showers...). This was a very popular demonstration at my high school teacher workshop. It is simple to do, although a bit messy.

Karen Meech
University of Hawaii

Simulating Ring Particles

To demonstrate the Voyager radio occultation experiment and how it was used to determine the properties of the particles in Saturn's various rings, I shine a laser through different materials. The laser is the pen-sized one I use as a slide pointer, but any classroom laser would do. Thin typing paper, wax paper, plastic wrap, and even nylon stocking fabric all work well as intervening media in that they all produce characteristic patterns on a screen when the laser shines through them. When I do this demonstration in the dark, I ask students to guess whether the pattern seen is produced by a tightly or loosely woven material. From this, I go on to talk about ring particle density and size distribution.

Thomas Hockey
University of Northern Iowa

Fluid Flow on Mars

Here is something that I recently stumbled upon. It is a demonstration of the fluid flow required to explain the ejecta blanket around certain Martian craters, such as the crater Yuty. These flows are given as evidence of subsurface water. A photograph of this crater in Figure 7.21b, p. 260 in Phil Flower's Understanding the Universe, 1990) looks very much like the results from a backyard experiment done after several inches of rain, when the sand was saturated. I believe that this last point is very important. I dumped a bucket of water on the sand which excavated a small crater and traveled across the surface as a slurry before the water managed to sink into the underlying saturated sand and left the ejected sand behind in a characteristic flow pattern.

I should emphasize that this discovery was accidental: I was merely emptying the contents of a nearly full bucket that I had used to measure rainfall so that I could measure more rain that was expected. I am sure that better results could be obtained using a garden hose to soak the sand.

Danny R. Faulkner
University of South Carolina at Lancaster

Planetary Bands

Put Ivory liquid (pearly) and water in a spherical glass bowl or flask. Rotate it, and see how it divides into bands like Jupiter, Saturn, and Neptune. See the Exploratorium Science Snackbook fluid exercise called the "Turbulent Orb". It discusses ways to study convection and turbulence using a fluid-filled sphere.

Elizabeth Roettger
DePaul University

Convection in Planetary Atmospheres

Convection in planetary atmospheres can be colorfully demonstrated by submerging a beaker (or small bottle) filled with warm water in a transparent container of cold water. The water in the small container is first darkly colored with food coloring and covered with plastic wrap. The plastic wrap is pierced with a sharp pencil after placing it at the bottom of the larger container. The larger container might be a small aquarium or any watertight clear container.

A demonstration of convection above a hot source can be performed in a simple manner by holding a lit candle in front of a slide projector. The projector beam is shown on a wall or screen beyond the candle revealing convection of the warm air and smoke from the candle flame.

Gene Maynard
Radford University

Jupiter-like Planets Around Other Stars

The direct detection of a planet beyond the solar system by imaging is extremely challenging. It is necessary to resolve the planet's reflected starlight from the overwhelming luminosity of the nearby star. For a Jupiter-Sun system observed at a distance of 10 parsec, this requires being able to detect a brightness ratio of 2 times 10^{-9} at visible wavelengths (or 10^{-4} at infrared wavelengths) just 0.5 arc seconds away from the star. The accuracy is currently unattainable, but can be achieved with space missions that have been planned or proposed.

A star and a planet orbit around a fixed center of mass. Motion about this common center produces a reflex motion or barely discernible "wobble" in the stellar trajectory. For example, the Sun wobbles around a point near its edge (or photosphere), and most of that motion is due to the influence of the most massive planet, Jupiter.

As in the example above, the maximum amplitude of the Jupiter-Sun reflex observed at a distance of 10 parsec is 0.0005 arc seconds. The astrometric accuracy that can be attained from the ground is a similar number, so Jupiter-sized planets can in principle be detected around the nearest stars. The same wobble induced by a planet orbiting a star can be observed spectroscopically if the plane of the orbits is nearly parallel to the line of sight.

In this case, the small change in relative velocity of the star as it moves toward or away from the observer can be seen as a periodic Doppler shift in the absorption line spectrum of the stellar photosphere. Jupiter's tug on the Sun causes a maximum wobble in velocity of 12 meters per second. Note that this must be detected on top of a star's net (or peculiar) motion through space of tens of thousands of meters per second! At distances of the nearest stars, planets of the size of Jupiter are marginally detectable with current technology. Using most of the techniques mentioned above, detection of Earth-sized planets is a hundred times more difficult.

Chris Impey
University of Arizona

Serendipity in Discoveries

Point out to students that many of the most important discoveries were not anticipated but were made by people making new observations with an open mind. For example, the rocky/icy satellites of the outer planets have turned out to be far more varied and unusual than anticipated.

In the Voyager flyby of Jupiter, as Voyager 1 was departing Jupiter it took some overexposed images of the satellite Io to use for navigation purposes. These images showed a large bubble-like object on the edge of Io--a volcanic plume of material. Other pictures of Io had been processed earlier to bring out surface detail. When these images were processed to enhance contrast, they also showed several volcanic plume features. The preoccupation with enhancing surface features did not encourage the discovery of plumes outside the surface.

Stephen M. Pompea
Pompea & Associates

Planetary Auroras and Nocturnal Light UFOs

A novel piece of equipment I use in connection with the solar system is a large, mostly evacuated, discharge tube to simulate planetary auroras and nocturnal light UFOs. I lecture in the college circuit that many nocturnal lights are caused by matter from comets entering the geomagnetic system.

Louis Winkler
Pennsylvania State University

Space Junk

To tether the idea of space junk (asteroids, meteors, comets) in students' minds, comment on the destructive capabilities of these objects hitting the Earth. For example, Barringer produced by a 50-ton or 50-meter rock (20 megatons), Tunguska by a 100,000-ton or 1-km rock (80,000,000 megatons), yet some asteroids are 100 km across. The shielding (and weathering) effect of atmosphere are indicated by comparing the Moon's and Earth's surfaces. Crude calculation shows that without protection, chances of a human being hit in a lifetime are roughly 10^{-9}. The way to reduce your odds is to sleep standing up, reducing the target area.

Chris Impey
University of Arizona

Galilean Moons of Jupiter

(For extra credit -- in-class quiz) Ask the students to develop a mnemonic for the four Galilean moons of Jupiter: Io, Europa, Ganymede, Callisto. They are given 2 - 3 minutes to develop a four word sentence -- no sentence fragments and no bad grammar, please! -- with initials I, E, G, C, in that order. I double the credit if they can come up with a sentence alluding to things or people astronomical. Some examples:

Is everybody going crazy?
Italians exalt Galileo's cleverness.
Is earth getting colder?
Intelligent earthlings give candy.
I eat green cheese.
Iowa exports great cows.

Rules:
1. No made-up proper nouns: names must refer to actual, well-known people or places or things, not fictional ones.
2. No auxiliary words (prepositions, conjunctions, etc.) allowed.

Dan Wilkins
University of Nebraska - Omaha

Cosmic Year

To get the feel of large timescales, use the cosmic year to describe the history of the Earth (4.5 billion years scaled to a year, with the present day at the end of the year). Can emphasize that although life formed as soon as it could on the cooling planet (March 20), civilization did not develop until December 31 at 11:59 P.M., and environmental modification did not start until 11:59:59.5 P.M., the last half second. Can be used as a lead-in to a discussion of environmental concerns, greenhouse effect, deforestation, pollution, etc.

Chris Impey
University of Arizona

Classifying Properties of the Solar System

Using drawn, purchased, or photocopied pictures of solar system objects, students must adopt an object and arrange themselves around the room according to various classification rules. The teacher provides the rule and students must arrange themselves without speaking to one another or dragging each other around the room. Both rank ordering and grouping classification schemes are useful.

Some Possible Classification Schemes

Rank Ordering	**Grouping**
closest to farthest from Sun	inner and outer
biggest to smallest	terrestrial and jovian
hottest to coldest	rings and no rings
low to high density	craters and no craters
number of Moons	moons and no moons

One variation is to have students draw enormous posters of their solar system object and write useful information to remember on the back before arriving at class. Another is to have a team of three students come up with as many different arrangements as possible, one at a time, arrange student-posters, and have class members try to "guess" their classification rules.

Tim Slater
Montana State University

Gravity and Weight on Distant Planets

TARGET CONCEPT: Gravity is the force that keeps planets in orbit around the sun and governs the rest of the motion in the solar system. Gravity alone holds us to the earth's surface and explains the phenomena of the tides.

Part I - Exploration: Is there any gravity on the Moon? How could you find out? As a first step, watch one or more movies of US astronauts working on the Moon.

Part II - Concept Introduction: How much do you weigh on distant planets? Which planet characteristics cause a planet to have more or less gravity? Consider the following variables--presence of an atmosphere, planet diameter, planet mass, planet temperature, and/or distance from the Sun. Which do you think is most important in determining a planet's gravitational strength? To investigate your hypothesis, find out how much you weigh on other planets using one of the many planetary-weight calculators available on the Internet. You might find it helpful to make a comparison chart.

Part III. - Concept Application: What causes a planet to have gravity? What will the gravity be like on the newly discovered planets around distant stars?

Consider each of the following hypotheses about what causes you to weigh more or less on other planets. Use an Internet-based Planetary Weight Calculator to prove or disprove the following proposed hypotheses.

Aaron's Hypothesis
Planets with thin or no atmosphere have little or no gravity—like Mercury.

Pat's Hypothesis
Planets that are cold have only a small amount of gravity—like Saturn.

Chris' Hypothesis
Planets that are massive and have the largest diameters have the most gravity.

Kesh's Hypothesis
Planets that have thick atmospheres and are farther from the Sun have the most gravity.

Assessment Strategy:
Write a brief paragraph that explains what characteristics cause a planet to have more or less gravity. Use examples from your research.

Tim Slater
Montana State University

Chapter 8 Physical Properties of Matter

Comparing the Electrical to the Gravitational Force

After calculating the ratio of the gravitational force between an electron and a proton to that of the electrical force, it is quite instructive to compare the gravitational attraction between Venus and the Earth to that of two men separated by the same distance but the one man has an excess of electrons by 1% and the other has an excess of protons of 1%. The two numbers come out to be about the same. This is always surprising to anyone that sees it and almost shocking.

V. Gordon Lind
Utah State University

Simulation of Proton-Proton Reaction

The purpose is to simulate the effects of the electrostatic repulsion of protons in the proton-proton cycle. The apparatus needed is a wooden stand with aluminum post; 2 ring magnets with Velcro strips attached. The electrical repulsion of the protons prevents their coming together close enough for the short range Yukawa force to be effective, unless the velocity of the protons is quite high -- corresponding to about 10^7 K in stellar interiors. (It would be even higher if quantum tunneling were not present, but that's another demo.) This simulation illustrates that with the repulsive force of the ring magnets mimicking the repulsive electric force of the protons and the binding effect of Velcro mimicking the short-range attractive Yukawa force. The speed of the magnet is derived from gravity free fall; the speed of the protons is induced by gravitational heating in the star.

Put the ring magnets on the stand with the Velcro ends face-to-face. This ensures that like poles face each other and the magnets repel. Lift the upper one from 1 to 3 cm above its equilibrium point and drop it. If you didn't lift it too high the two magnets won't join. Now lift the upper magnet about 5 cm and drop it. The Velcro ends will join and if these were protons a deuteron would result.

To simulate the CNO cycle reaction put 6 magnets on the bottom and note that the dropped magnet now must be dropped from a higher height. This mimics the higher ignition temperature of the CNO cycle owing to the greater repulsion of the incident proton by the carbon nucleus. To more advanced classes point out that the analogy is simplistic because tunneling is required for computing the correct ignition temperature and burning rates. There is no easy way to incorporate quantum tunneling into the simulation.

John Broderick
Virginia Polytechnic Institute and State University

Infrared Camera Demonstrations

Near infrared camera are becoming widely available, and they can often be adapted with a relay lens for use in the classroom. The scanned video output needs to be fed to a monitor or projection television for use in a large classroom. A series of simple experiments can graphically demonstrate the nature of electromagnetic waves beyond the visible spectrum. Imaging a person's face to begin with shows the hot and cold regions, and the heat loss features like the nose and ears can be pointed out. Imaging a soldering iron with the lights on and then off shows that the camera is clearly seeing something that is not visible (it will also show how poorly designed most soldering irons are!). Insert a black cloth between the soldering iron and the camera to show that it does not stop the infrared radiation. It is also effective to image the soldering iron by reflection in a dull metal plate. Although the surface is not good enough to cleanly reflect visible light, the larger infrared waves see a much smoother surface. A blow dryer can be used to "draw" on a metal door or cabinet, with amusing results. Conversely, you can use an ice cube to "draw" cold regions on a volunteer's arm or face. With this setup, you are only limited by your imagination!

Chris Impey
University of Arizona

Nonthermal Radiation

Contrast the thermal motions (and spectra) of atoms with the nonthermal radiation produced by cosmic accelerators. The Superconducting Supercollider will give 10^{12} electron volt protons for 5-7 billion dollars, our own galaxy produces 10^{15} electron volt protons for free. Point out that if a grain of sand entered our atmosphere with the same energy per nucleon as a 10^{15} electron volt proton, it would be the equivalent of a billion tons of TNT.

Chris Impey
University of Arizona

Compression of Gases and Their Temperature

A dramatic way to illustrate the rise in temperature of a gas that is being compressed (e.g., a contracting protostar) has been used in some physics classes at UC Berkeley, and I have found it fun to do in my introductory astronomy course. Get a heavy-walled glass cylinder with a closely fitting piston. Drop a small wad of cotton soaked with ether into the cylinder, letting it fall to the bottom. Insert the piston and drive it down into the tube with a very sudden, forceful push. If done successfully, the cotton should ignite, producing a bright flash. It does this because the compression substantially raises the temperature of the gas.

Alex Filippenko
University of California, Berkeley

Conservation of Energy

As an example of the transformation and conservation of energy, take the act of listening to a singer on the radio. A mechanical vibration in the singer's larynx produces a sound wave, which is converted into a mechanical vibration in a microphone, and then into a modulated electrical signal. This modulated signal is amplified and filtered and then generates accelerating electrical charge in a transmitter, which leads to an electromagnetic wave. The process then continues in reverse order to produce the sound wave that hits your ear.

Chris Impey
University of Arizona

Why Is the Sky Blue: Scattering Phenomena

A good demonstration to illustrate that blue light is scattered more strongly than red light by small particles is to put a few drops of milk in a beaker or glass. Then shine a light (a flashlight works well) through the beaker. In fact, putting the glass on top of an overhead projector works great, also. The blue light is preferentially scattered out of the light beam, and the liquid looks blue. Cigarette smoke looks blue, for the same reason, when it is illuminated by a bright light. The light beam that goes through the liquid is now redder or more orange in color, since the blue light has been scattered out of the beam. The reddish color is similar to what happens at sunset, when the greater path length of sunlight through the atmosphere scatters out most of the blue and leaves the red. Blue light is scattered strongly because the particles (the colloidal milk particles) are similar in size to the wavelength of blue light. As a general rule of thumb, the greatest amount of scatter occurs when the particle size is equal to the wavelength.

The actual reasons why the sky is blue are a bit more complicated, but this demonstration is pretty, easier to use than the sulfur one often found in experiment books, and illustrates the principle quite well.

Stephen M. Pompea
Pompea & Associates

Matter

Crudely demonstrate that atoms and molecules are much smaller than the eye can see (and that molecules have shape) with this experiment. Fill a shallow glass-bottomed container with water and sprinkle pepper on the surface of the water. Put container on overhead projector, so class can see pepper. Drop one drop of dishwashing liquid on surface of water; pepper will rush away as dishwashing liquid tries to form monolayer. Quick calculation on board ($4/3\ \pi r^3 = \pi r^2 h$) will show that the molecules are less than 10^{-4} millimeters, or one thousandth of the thickness of a human hair. Stress that all students are 99.99999999999% empty space, but even though the electrical force is far stronger than gravity, the near perfect neutrality of matter causes gravity to dominate on the large scale.

Chris Impey
University of Arizona

What Is an Erg?

An erg is a unit of energy. It is a fundamental unit in the cgs system.

But what is it? If you tell your students it is a dyne acting over a distance of one centimeter or that it is 10^{-7} joules, you may lose them. The output of a 100-watt light bulb is a billion ergs per second, but this is still hard to picture. An erg is not as easy to picture as a second or a gram. How can an erg be visualized? I'll never forget the image created in a lecture by Roger Culver who said:

> An erg is the energy a mosquito uses to do a pushup.

You can calculate for yourself whether this is indeed correct or more of an illustration of the insignificance of one erg. In any case, an erg is a very small unit of energy.

Stephen M. Pompea
Pompea & Associates

Interferometry

When introducing interferometry, relate it to a football stadium public-address system. The sound arrives at different times due to differences in travel time from the different speakers. Interferometry is the reverse process; the signal, electromagnetic radiation, leaves the source at the same time, but arrives at different spots on earth at different times.

Stephen M. Pompea
Pompea & Associates

A Model of a Hydrogen Molecule

Materials:
An 18-inch wood or aluminum rod0.25 inch in diameter. About 25 inches of bare copper wire. Two red Ping-Pong balls.

Use any round handle (0.5"- 1" in diameter) to wrap the copper wire to form a spring with 10 - 12 coils. Leave about 1.75" straight wire sections on each end of the spring. Drill holes (with high-speed drill) through opposite sides of Ping-Pong ball, and insert a ball on each side of the spring, bending the protruding end of the wire so that each ball is held in place. Cut a slot in the tip of the 0.25 inch rod so that a portion of the wire in the middle of the spring will fit tightly into the slot. The rod simply allows one to hold up the device. Each red ball represents a hydrogen atom. The spring demonstrates the vibration modes available in the molecule, and spinning the rod in your hand demonstrates the rotational modes of the molecule.

Richard D. Schwartz
University of Missouri - St. Louis

Doppler Shift

Probably every astronomy teacher uses the illustration of a train whistle pitch change to introduce the Doppler shift. Many students today have more familiarity with race cars, such as in the Indy 500. So, if you can do the sound effects of a race car coming toward you and then past, use that as an introduction.

An easy way to demonstrate the pitch change is to use a piece of surgical tubing attached to a whistle. Then whirl the whistle around your head, while blowing it. The pitch will be heard to change by the student as the whistle approaches and recedes from them. Have the students note when the pitch is highest, lowest, and intermediate. Extend this analogy to Doppler radar measurements of the rotation period of Venus and Mercury, or to the radial velocities of double stars.

Stephen M. Pompea
Pompea & Associates

The Difference Between Quantized and Continuous

To help students understand the difference between quantized and continuous, which I think is essential in the study of stellar spectra, I use examples like sibling, money and pregnancy for quantum things and height, weight, and age for continuous things. Questions like "Does anyone have 2.1 brothers? Could you be 20% pregnant? Has anyone here ever been 4 feet tall?" generally draw laughs yet point out the difference between the two types of distributions.

Karl Giberson
Eastern Nazarene College

Understanding Radioactive Decay

Astronomers teach radioactive decay in relation to supernova remnants, dating of meteorites, the age of the earth, and other areas. It is also one of those fundamental physical concepts that we hope to teach our students since they may be taking their first and last college physical science course. However, the concepts involved in radioactive decay are hard ones to get across. The following demonstration may help illustrate the randomness of the process and the variation in half-life.

Radioactive decay is a little like making popcorn. You never know when any individual kernel of popcorn is going to pop. When it does it pop it is irreversibly changed. You can say, however, that after a certain amount of time (depending on temperature and other conditions) most of the popcorn will have popped.

Like popcorn, you cannot predict when a particular nuclei of a radioactive species will decay (pop). However there is a rate associated with the overall conversion process. Unlike popcorn, the rate does not depend on the gross physical conditions. Changing the temperature does not affect the decay rate. The half-life is the time for half of the kernels to pop. How long will it take for all the kernels to pop? Like popcorn, it takes a very long time for all of the nuclei to "explode." In the case of radioactive decay: Ask the students how long it would take, if two radioactive atoms are left, for one to explode. The answer, as always, is another half-life.

Half-lives vary tremendously. For example, one state of ^{170}Yb has a half-life of 1.57 nanoseconds while the half-life of ^{37}Ar (the radioactive species counted in neutrino detection experiments) which decays to ^{37}Cl is 35 days. An even longer lived isotope is rubidium; ^{87}Rb decays to Strontium ^{87}Sr with a half-life of 46 X 10^9 years. The popcorn concept helps many students who have no experience of radioactivity grasp the essential concepts.

Stephen M. Pompea
Pompea & Associates

Hydrogen

Hydrogen is mentioned so frequently in astronomy that I find it useful for students to see this element prepared with hydrochloric acid and zinc and then used to inflate soap bubbles.

Gene Maynard
Radford University

The Photoelectric Effect and Solar Cells

Strangely enough, Einstein won the Nobel Prize, not for his theories of relativity, but for the photoelectric property of light. An electroscope is a device for storing electric charge. It consists of an evacuated glass bulb through which a metal rod protrudes. The rod has a hook on the end across which a thin strip of gold is laid.

When electric charge, say from a rubber comb, is touched to the outside tip of the rod, electric charge travels down the rod and builds up on both sides of the gold strip. Since the charge is all negative (or positive) the two sides of the strip repel each other. Now when the charged electroscope is taken out into sunlight, it quickly loses its charge. Einstein proposed, correctly, that particles of light actually knock the charge off the gold, eventually making it lose its charge.

Solar cells make use of the photoelectric effect to make electricity. Obtain a small solar cell and an inexpensive voltmeter. Connect the two leads from the solar cell to the voltage inputs of the voltmeter. Now shine a flashlight on the solar cell. The voltage reading will shoot-up. When the flashlight is removed, the voltmeter reading will return to a small value. What happened? Solar cells are made of wafers of either silicon or gallium arsenide. Normally when wires are connected to either end of these semiconductors, no current will flow through them. But when light strikes the semiconductor, it knocks off electrons. These electrons then flow through the wire. The more intense the light, the more electrons are knocked off and the greater the observed voltage.

Constance Walker
University of Arizona

Liquid Nitrogen Demo

This is always a favorite demonstration for my students. You can demonstrate the volume difference between liquids and solids. Also you can demonstrate the boiling points of different elements such as liquid O_2, N_2, and H_2O. Discuss with students how materials change their properties at LN_2 temperatures. They can consider this for the design of a space vehicle which holds liquid nitrogen. Ask students to make predictions and then try dipping things in liquid nitrogen. Some things to try include rubber tubing, superballs, superconductors, fruit, and a raw egg.

Constance Walker
University of Arizona

Chapter 9
Light and Electromagnetic Radiation

Emission and Absorption of Photons

Here is a good way to demonstrate what happens when an atom emits or absorbs photons having different energies, as well as the fact that not all photons can be absorbed. It also gives you a chance to exercise while lecturing -- but be careful not to hurt yourself! Set up two chairs of different heights and unequal spacings, and a large table that is even higher; I generally use a chair about two feet above the floor, a chair about a foot higher, and a slightly higher table. (Or, use the stairs in the lecture hall, if available.)

The chairs represent the discrete bound energy levels of an atom or ion, and the table represents the continuum of unbound energies. Get some red, yellow, green, blue, and violet balls (e.g., juggling balls). Each ball is a photon, with color corresponding to the wavelength (and hence energy) of the photon. Give the violet, blue, and yellow balls to a student, keeping the green and red balls for yourself. Explain that you are an electron bound to an idealized atom or ion (which can be represented by a nearby basketball), that the floor and two chairs are your three possible bound energy levels, and that the table represents unbound energy levels.

Stand on the floor, and ask a student to throw you the blue ball. As you catch it, jump from the floor (first energy level) up to the third energy level. This represents the absorption of a high-energy photon. Jump down to the second level, releasing the red ball (not necessarily towards the student). Then jump down to the floor (first level), releasing the green ball. (Here I assume that the height difference between the floor and second level is larger than that between the second and third levels.) Besides illustrating the quantization of energy levels in atoms and the role of photons in emission and absorption, this demonstration shows that a single high-energy photon can be converted into two or more lower-energy photons.

Now ask the student to throw you the yellow ball. Do not jump when it goes past you or hits you; the energy of the yellow photon does not correspond to the difference between any of your hypothetical energy levels. Finally, ask the student to throw you the violet ball. When you catch it, hop up on the table and start wandering around, to indicate that you have become a free electron, not bound to the atom (illustrating ionization). If you wish, this entire exercise can be modified so that students jump from level to level while balls are thrown at them or by them. It gets students actively involved while learning some important physics.

Alex Filippenko
University of California, Berkeley

A Classroom White Light Source

While a diffraction grating makes a fine spectrum, I like to use the more familiar prism when first demonstrating dispersion in front of the class. The brightest source of nearly white light in the classroom is usually a slide projector. I cut an aperture out of an expendable thirty-five millimeter slide and place the slide into the projector's light path; this forces the light path to remain open. When the projector is turned on, a white rectangle appears on the screen at the front of the room. I then put a large glass prism into the beam. Students can see that light has been refracted out of the beam because a dark shadow now appears in the rectangle. Where is the "missing" light? It is spread out into a spectrum, which, with slight changes in the orientation of the prism, I can direct onto a side wall, the ceiling, or dozing students.

Thomas Hockey
University of Northern Iowa

Production of Emission Lines in HII Regions

HII regions produce emission lines in several ways. The two most important are (a) ionization by absorption of an ultraviolet photon, followed by recombination and cascading through lower energy levels; and (b) collisional excitation or ionization by a free electron, followed by recombination (in the latter case) and cascading. To illustrate these concepts, you will need the materials described in the previous demonstration entitled "emission and absorption of photons; ionization." Give a student (a hot star) the violet ball, and keep the others.

For (a), consider yourself to be an electron "orbiting" a positively charged nucleus (which can be a basketball on the floor) by walking around the nucleus in circles. Ask the student (hot star) to throw you the violet ball. When you catch it, hop up to the top of the table and start wandering around in various directions. This is ionization, and you are now a free electron. Gradually make your way back to a basketball (not necessarily the one from which you came, but one next to the two chairs).

As you release a yellow ball, jump from the table top to the upper chair (third energy level); this is recombination. Then jump either directly down to the floor (first) level while releasing a blue ball, or to the second level (releasing a red ball) followed by the floor level (releasing a green ball). Thus, a high-energy photon (violet ball) from a hot star has been transformed into several lower-energy photons.

For (b), once again consider yourself to be an electron "orbiting" a positively charged nucleus. Ask a student (now playing the part of an energetic free electron) to kick you. Either hop up to the table (collisional ionization) and follow the steps given above, or hop up to a chair (collisional excitation) and cascade down, releasing the appropriate balls along the way. Besides illustrating how HII regions work, this demonstration provides students a chance to vent their frustrations by kicking the lecturer.

Alex Filippenko
University of California, Berkeley

Using Filters and Diffraction Gratings

Give each student a set of color filters (red, green, and blue - can be bought from Edmund Scientific or scholastic supply houses, or buy 'gels' from a theatrical supply house, and tape them behind holes cut in index cards). Be sure the blue filter does a good job of filtering out red & yellow light. If possible, also give them a diffraction grating. Do the usual activities - look at different colored objects, produce a rainbow and look at it through the filters, place colored objects in the rainbow and compare to white light/filter combinations.

Now project color slides of astronomical objects and look at them through the filters. Be sure to use a few slides of nebulae and star-forming regions. With good filters, the slides will look very different through the various filters. You might be able to see dust in one filter and stars in another, for example. This recreates the science astronomers get from broad-band (filter) imaging techniques, different wavelengths (colors) show different things in the same space.

It's not quite the same, since you're looking at pictures that have been separated into three colors and recombined so they'll look right to human eyes - much as shining three colored beams onto the same spot can make the spot look white to human eyes (Look at it through a diffraction grating and you'll see colored bands instead of a whole spectrum.). Still, if the filters you're using are a reasonable match to the ones the slide-maker used, it will be a good approximation of broad-band imaging, the essential science will be preserved, and the effect is quite striking. (This comes from a workshop by Edna DeVore, of the FOSTER project.)

Elizabeth Roettger
DePaul University

The Doppler Effect

There are a number of good ways to demonstrate the Doppler effect. Among them: 1) Ripple tank film loops or videotapes showing the effect in water waves; 2) Audio or video tapes of cars passing by at various speeds; 3) A whistle that mounts on a rotator so that it moves toward and away from the listener; 4) A battery-powered whistle that can be inserted in a foam and a rubber ball and then thrown around the room. (You might videotape the cars from an ESPN auto race; the rotating whistle can be purchased from scientific suppliers such as Sargent-Welch -- about $70 -- and a whistle is available at Radio Shack which works well in a foam ball.)

Karl Kuhn
Eastern Kentucky University

Simple Diffraction Demonstration

A single-filament incandescent lamp (sometimes called a showcase light) can be used to show diffraction. Cut slits in index cards, pass them out to the students, and tell them to look through the slits at the lamp as they flex the slits open and closed. (They can even look through a small opening formed between their fingers, but the cards work better.)

Karl Kuhn
Eastern Kentucky University

Spectra

To give a simple demonstration of spectra, obtain some inexpensive plastic diffraction grating (supplied in 2 x 2-inch slide mounts), pass them out to the class and let them view an incandescent lamp and a few vapor lamps. The "ooh's" and "aah's" heard when students view the spectrum for the first time is a pleasant sound. I have found that this works in lecture halls with as many as 100 students. Most gratings are returned, and perhaps those that are not returned will be put to good use viewing street lights and other scientific phenomena.

Gratings in 2 x 2-inch slide mounts are available from Edmund Scientific Company (101 E. Gloucester Pike, Barrington, NJ 08007) for about 50 cents each. A less expensive method is to buy a roll of diffraction grating film from the same source and cut it up yourself. If you do this, you might staple pieces of it over holes which have been punched into pieces of card stock -- perhaps 1 x 3 inches in size. Assuming low-cost help to cut and staple the gratings, these are inexpensive enough that they can be given to students.

Karl Kuhn
Eastern Kentucky University

Wave Nature of Light

When discussing the wave nature of light, hand out a number of sets of hand-held Polaroid "lenses." Have the student hold one Polaroid fixed and rotate a second one while viewing a light. Have them observe the alternate lightening and darkening of the light and ask them how many times the light grows dim and then bright during one rotation of the second Polaroid. There will be two "epochs" of each. The discussion can then proceed to the idea of a rope tied to a wall and passed through the slats of two pieces of crating material. When the openings of the two pieces of crate are parallel to each other, it is possible to produce waves in the rope that will pass through both crate sections. When the crate sections are at right angles to each other, the waves are blocked. If one crate is rotated a full 360°, there will be two places where the waves will pass through ("brightening") and two places where the waves will be blocked ("darkening").

Place a piece of diffraction grating over the lens of a standard 35-mm. camera and photograph a variety of elements in discharge tubes. Photograph the spectra from a camera mounted on a tripod. Photograph all spectra from the same physical setup. Using 400 ISO film with relatively short exposures (1/15 second) will produce a set of projection spectra. Use a spectrum of hydrogen or helium to calibrate a projection system. Project the slide onto a meter stick taped to a blackboard. Move the projector to a position such that the known lines of the calibration slide fall on the meter stick. One mm. will equal 30 Angstroms. Project other spectra and have students read to values for the wavelengths for other common elements and complete a spectral identification lab in a lecture setting. For added interest, suggest that students use a similar Polaroid arrangement to photograph street lights, room lights, LEDs, etc.

Darrel Hoff
Luther College

Depolarization of Polarized Light by Multiple Scattering

An excellent demonstration of how multiple scattering can depolarize linearly polarized light can be done using crossed polarizers and an overhead projector. A piece of wax paper is put between the crossed polarizers and the polarizers are laid on top of the overhead projector. The wax paper appears lighter as it has depolarized the light. The light is scattered many times inside the waxed paper and the light emerging from the wax paper is completely depolarized.

Stephen M. Pompea
Pompea & Associates

Why the Sky Is Blue and the Sunset Is Red

The sky appears blue because the particles in the atmosphere scatter the sunlight incident on the atmosphere. The molecules of air and the smallest particles in the atmosphere scatter the shorter wavelength blue light more effectively than the longer wavelength red light. A beam of sunlight loses more of its bluer colors as it penetrates the atmosphere. The scattered blue light causes the sky to appear blue. At sunrise and sunset, the sun's rays travel through a longer path in the atmosphere and enough blue colors are filtered out to cause the reddish glow in the sky. If the sky did not scatter any of the sun's light, it would appear black. The length of the wavelength of light and the size of the scattering particles is the critical physical cause of the color of the sky. On Venus where nearly all the light is scattered by the atmosphere, only some of the redder visible light penetrates, giving the sky a reddish color.

I demonstrate this effect by passing light through a solution with different size particles in suspension and observing the results. A water solution containing a mixture of hypo and dilute sulfuric acid will form a colloidal suspension of growing particles. As time passes, the solution changes from completely transparent (particles are too small), to a blue haze (scattering blue colors), to a blue-white haze, to opaque (all light is scattered). I use a Kodak slide projector and pass the beam through a two-gallon aquarium and then project it onto a screen. Add to the water 50 cc. of hypo (photographic fixer) and 20 cc. of dilute sulfuric acid; mix well. Be sure to perform a trial run to determine the length of time needed to complete the demonstration. The time for the required changes in particle size depends on the concentration of the reactants. As the particles grow, the blue sky appears in the solution and white light projected on the screen turns into an orange sunset color. I generally do this demonstration along with my demonstration of the bright line spectra of different gases since both require a darkened classroom.

Joseph W. Mast
Eastern Mennonite College

Light

Use boxcar analogy for relationship between frequency and wavelength (trains with different length boxcars traveling at same velocity). Contrast subjective (blue is cool, red is hot) with objective color. Light is only one type of EM wave, but note that its importance stems from adaptation of our most sophisticated sense to the atmospheric window.

Note that creatures in other environments (some deep-sea fish, bats at night) use other senses (infrared, sonar). Can demonstrate wave/particle nature of light with a laser and Astroscan telescope. Compare laser projected directly on wall, with laser after projected through eyepiece of Astroscan (where, after multiple reflections, it will show interference fringes).

Chris Impey
University of Arizona

Thermal Spectrum

Convey in everyday terms the principle of a thermal spectrum, and the relationship between wavelength of emission and temperature. As you heat up an iron bar, it glows dull red, then orange, then white, but point out that even at room temperature, it emits a thermal spectrum. Infrared radiation can be demonstrated with a small infrared camera (commercially available) and a soldering iron. Another demonstration of the reduction of molecular motion as the temperature falls is to dip a rose (or racquetball) in liquid nitrogen and smash it. As a classroom aid, can tape the diffraction grating slides that are cheaply available to the ends of the cardboard tubes, with a slit at the other end, and have students look at the incandescent lights in the classroom.

Chris Impey
University of Arizona

Blackbody Radiation

To demonstrate the color and relative intensity of blackbody radiation as a function of temperature, I use a variable transformer to supply power to a large, clear bulb (a large projection bulb as used in theater lights or any high-wattage clear bulb with a filament that can be seen in class). I can demonstrate radiation from an object that is below 3000K. With an increase in voltage, the filament first emits heat and then has a dull red glow as the filament reaches 3000K and looks yellow as 5800K is reached.

The filament glows brightest and more blue as the voltage is quickly raised above 120 volts and quickly lowered before the filament melts.

Danny Overcash
Lenoir-Rhyne College

Doppler Effect

A small battery-operated smoke detector may be moved rapidly in a circle with the sound activated by pressing a button. This will produce a very obvious Doppler effect which is helpful in teaching about spectroscopic binary stars and other orbital notions which produce Doppler shifts.

Paul Nyhuis
Inver Hills Community College

Inverse Square Law

A slide projector is used to project a rectangle of light on the chalkboard. This area is outlined with chalk and brightness noted. The projector is then moved twice as far away from the blackboard and the area marked again. This is a useful demonstration of the inverse square law of light or any inverse square rule.

Gene Maynard
Radford University

Refraction

To understand why light bends when it travels from one medium to another, imagine yourself looking for a short cut on your way walking home from school. On the way, you must cross a large open field and a thickly wooded forest full of briars.

Which route will get you home quickest (so as not to miss the Flintstones on TV)? You could take the shortest path to the woods, but then you would have to slowly work your way through a great distance of brush. On the other hand, you could take the longest path and run quickly through the open field to take the shortest path through the woods, but the increase in your total distance traveled would more than offset any gain in your average speed. What about a straight line? If you took a straight line route you would travel the shortest total distance, but since you travel more slowly through the woods you would still end up wasting valuable time. The quickest way home is to compromise and take a somewhat longer path through the field than through the woods, while not increasing your total path length too much.

Thus, it is a tradeoff between the speed you are able to travel over a given terrain and how far you must go in that terrain. This "minimum time" path is the same path that a photon of light would take in going from one medium of a given refractive index to another. The resulting relation between the angles and velocities of light in the media is simply Snell's law. If you are able to travel at the same speed in both media, the shortest travel time between any two points is along a straight line, as expected.

Glenn M. Spiczac
Bartol Research Institute

Electricity, Magnetism, and Light

Students often lack even the most basic understanding of electricity, magnetism, and light. Some simple explanations and experiments can be of great value to them. Electrons moving in a circle generate a magnetic field. Wind some copper wire on a nail. The greater the number of turns per inch, the greater the number of electrons in motion per inch, and the more intense the magnetic field density. Scrape the insulation off the ends of the wire and attach them to the terminals of a battery. Measure the field strength by picking up paper clips. How many additional turns are needed to pick up an additional paper clip?

A spinning magnet can "pull" electrons around a coil of wire. These moving electrons constitute an electric current and can be used to turn-on a small light bulb. This phenomenon can be demonstrated with an inexpensive bicycle generator and lamp. Monitor the intensity of the light bulb as a function of rotation speed.

Put a small transistor radio in an airtight (see-through) canister. With the radio on, pump out the air in the canister. Most of the sound will go away. However, some will remain. Why?

In contrast to sound, photons (electromagnetic radiation) can travel through empty space. Discuss with your students why this can happen. Discuss how the electric and magnetic fields keep each other going!

Constance Walker
University of Arizona

Diffraction and Interference

For discussing interference and diffraction, Robert Ehrlich in his wonderful book, *Turning the World Inside Out,* and 174 other simple physics demonstrations (Princeton University Press), suggest using a template consisting of concentric circles. This template can be used to illustrate waves from a point source. By duplicating this onto two transparencies, you can overlay them and change the spacing to illustrate interference and diffraction effects on an overhead projector. The superposition of these transparencies is a powerful way to demonstrate these phenomena and easier to use than a ripple tank.

Stephen M. Pompea
Pompea & Associates

Demonstrating Wave Motion

Half-fill a Pyrex baking dish with water and place it on top of an overhead projector. Put a small obstacle (e.g. a salt shaker) in the dish. Use the tip of a pencil to generate waves from one side of the dish. Note how the waves interact with the obstacle. Huygens imagined that light traveled through space and casts shadows in much the same way.

Constance Walker
University of Arizona

Making Rainbows

I first tried this series of demonstrations with my son's pre-school, but it has applications to all levels. The object is to get the students to understand the formation of rainbows.

First, start with a fish tank of water in a dark room. Shine a flashlight through the tank to show the colorful patterns that can appear on the wall. This demonstrates that water can split white light into a spectrum of colors. Then explain that this is how rainbows are made. However, instead of a bunch of fish tanks floating in the sky (the little kids think that this is funny) there are millions and millions of tiny droplets. Banging out lots of dots on a chalkboard with colored chalk tends to hammer this point home!

Next, to demonstrate this, give each student a spray bottle full of water and head outside on a clear, sunny, and calm day. Have the students put the sun behind them and have them spray in a circle around the shadow of their head. This should produce rainbows, and illustrates the fact that you must put the sun behind you to see a rainbow.

You can extend this exercise a little bit by asking students to try and see other students' rainbows. Sometimes they can't, and this shows that each individual sees his or her own personal rainbow! The success of this exercise was shown when the pre-schoolers sent me a thank you note, and they all drew their rainbows with a bunch of colored dots!

Jeff Yuhas
Reading Memorial High School

Color of Stars

Use a Variac and an incandescent lamp to show changes in color with increasing voltage. Compare the colors to the temperatures of stars. On a clear night, have your students look at a few of the brightest stars. They may find that they appear reddish or bluish. Our own Sun appears yellow. Have your students discuss what they would conclude (based on their color alone) about the relative temperatures of these stars.

Constance Walker
University of Arizona

Chapter 10
Telescopes, Optics, and Light Pollution

Radio Telescopes

Radio Telescopes pick up radio waves from outer space and hence are really antennas similar to the satellite dishes found in many localities for TV reception. A crude but effective demonstration of radio signal pickup can be done using a portable radio and a dry cell and a length of insulated conducting wire. Turn on the radio so that an audible signal is heard when the frequency is tuned between stations. Usually only background hiss is heard. Now with the dry cell nearby, connect the wire to one pole of the cell and briefly touch the other end of the wire to the other pole, momentarily shorting out the cell. Audible static will be heard on the radio.

Repeated contact on the cell will result in noise at radio frequencies as evidenced by reception on the radio. Rubbing the wire against the contact is even better. If the cell is slowly moved around the radio, or farther away, the signal will decrease in intensity indicating change in location. Similar behavior is used to locate radio sources in the sky.

James E. Kettler
Ohio University - Eastern Campus

A Demonstration of the Resolving Power of a Telescope

To demonstrate graphically that resolving power depends on the aperture of a telescope, prepare a card with a narrow slot that is a mm or two wide, and as long as the aperture of the telescope. (This also can be done with two cards and a pair of binoculars.) Fix the card in front of the telescope or binocular objective(s) with the slot(s) vertical, and have the students look at a brick wall in the middle distance, such that individual bricks are visible. They should be able to see the horizontal mortar joints but not the vertical ones. Then place the cards such that the slots are horizontal: the vertical joints should now appear, and the horizontal ones disappear. More advanced students can relate what they see to the spatial frequencies passed by the slot in the two orthogonal directions.

William G. Weller
Association of Universities for Research in Astronomy

An Experiment to Show That Telescopes Do Not Give Images of Stars

Stars are so distant that they subtend at most a few milli-arcseconds. No telescope in existence is capable of showing their surfaces, but rather forms an Airy disc, i.e., the two-dimensional Fourier Transform of a circular aperture, illuminated uniformly by starlight. To demonstrate this, cut a series of apertures in a card which is just a little larger than the aperture of the telescope. These could be square, triangular, and elliptical, for example. Mount the card in front of the telescope, and view bright stars at high magnification.

If the "seeing" is fairly good, and the aperture not too large, the diffraction image of the various apertures should be apparent, and they should be quite different from the image seen in the unmodified telescope. To show that this is an effect for unresolved (point like) sources only, use the same apertures to view the moon and planets, and little if any effect should be visible.

William G. Weller
Association of Universities for Research in Astronomy

Increase in Sensitivity

Let students observe each other's pupil diameters in light. Turn off light for short time and observe diameters immediately after light is turned on. You might also have very dim objects to show increase in eye sensitivity with time (2000 x in 20-30 minutes)

Neal P. Rowell
University of South Alabama

Telescope vs. Eye

The naked eye can see stars down to 6th magnitude, the largest telescope can see faint galaxies down to 30th magnitude. The difference is a factor of 4,000,000,000; where does it come from? From a difference in aperture (1/2 inch vs. 6 meters, or a factor of 500), a difference in detecting efficiency (0.1% of all photons for the eye vs. 50% for a CCD, or a factor of 500), and the difference in integration time (the eye reads out every 1/10 second vs. a 30-minute integration with a CCD, or a factor of 16,000).

Chris Impey
University of Arizona

Large-Scale Structure

A set of slides that is very effective in showing the enormous mass content of the universe is three that show a small patch of sky, a) from a Palomar sky-survey plate, b) from a 4m photograph, and c) from a long exposure CCD frame. The patch is only the size of the head of a pin held at arm's length, but in the last slide the frame is filled with thousands of galaxies (can lead in to Olber's Paradox here). Multiplying up shows the many millions of galaxies, each with billions of stars.

Chris Impey
University of Arizona

The Mars Illusion

Why did so many observers see canals on the planet? We now know that disjointed surface features at the limit of visibility were joined into canals by the human brain.

This demonstration gives students an experience of how this happens. Make a small drawing of Mars or, better yet photocopy and color an old drawing of Mars that shows features, but not any obvious canals. Now set up a very small telescope (1" aperture is fine) or binoculars on a test tube stand. Put the telescope slightly out of focus if it is too good a telescope (or put the drawing farther away, to simulate how Mars would look at a good opposition and have students draw what they see. Use a penny to draw the circle for the disk of Mars so students have a starting place on their drawing. A significant number of students will join together surface features to make canals. This is also an excellent time to discuss optical illusions and the different roles of photography and optical observations, and the role of atmospheric seeing on image quality.

Stephen M. Pompea
Pompea & Associates

Resolution

To demonstrate resolution, converge two helium-neon laser beams on white paper taped on the blackboard. Next, slowly separate the beams to show two distinct laser beams. Also may be used to illustrate visual binaries.

Michael Broyles
Collin County Community College

The Cheapest and Smallest Newtonian Reflector

I made this reflector when I was young and still recommend it as a simple illustration of a Newtonian Reflector. It is crude, but it works, and is made from common materials. The mirror is a lens from an eyeglass. Almost any lens will do, as long as there is a concave surface. Paint the back side black. Determine the focal length of this mirror by reflecting sunlight off of it. A toilet-paper tube will act as the tube of the scope. A pipe cleaner can be the diagonal mount, and a piece of a second surface mirror can be attached to it as a secondary. A hole in the tube can hold a little eyepiece or eyepiece lens. Any simple lens will do, though some experimentation is necessary. If no lens is available, it can be used at prime focus, with the prime focus being in the usual place for an eyepiece in a Newtonian system. For a more sophisticated design, aluminize the eyeglass lens, and use a small piece of first surface mirror as the diagonal. The telescope works amazingly well for objects like the moon, even if the primary is not aluminized.

Stephen M. Pompea
Pompea & Associates

Radio Jove Project

Radio Jove is an educational project to involve secondary school students in collecting and analyzing observations of the natural radio emissions of the planet Jupiter and the Sun. Participating students get hands-on experience in gathering and working with space science data. The project is a joint effort of NASA's Space Science Data Operations Office, the University of Florida Astronomy

Department, the Florida Space Grant Consortium, high school teachers, and volunteers.

Radio JOVE is an interactive educational activity that brings the radio sounds of Jupiter and the Sun to students, teachers, and the general public. This is accomplished through the construction of a simple radio telescope kit and the use of a real time radio observatory on the Internet. The Web site contains science information, instruction manuals observing guides and education resources for students and teachers. The target audiences are high school and college science classes with some activities being extended into middle school classrooms.

Goals of the Radio Jove Project
--Educate people about planetary and solar radio astronomy space physics and the scientific method
--Provide teachers and students with a hands-on radio astronomy exercise as a science curriculum support activity by building and using a simple radio telescope receiver/antenna kit
--Create an on-line radio observatory that provides real-time data for those with Internet access
--Allow interactions among participating schools by facilitating exchanges of ideas, data and observing experiences

The combined receiver and antenna kit costs $100. For more information visit the Website at
(The above information was excerpted from their web site)

http://radiojove.gsfc.nasa.gov

Stephen M. Pompea
Pompea & Associates

Why Can You Receive AM Radio Better at Night?

The ionosphere is not nearly as familiar a layer of the atmosphere as the troposphere or the stratosphere. To introduce the ionosphere to students, I begin by asking the following question: Why can you receive AM radio better at night than during the day? This question starts as the jumping off point into an exploration of the principles of radio transmissions, atomic structures, electromagnetic waves, the ionosphere, and state-of-the-art remote sensing capabilities.

To start the lesson, I have students take AM radios home to notice the farthest stations they can receive at night and then have them try to pull in those stations during the day. When they can't, this opens up a discussion of "Why Not?" We then discuss the fact that radio waves can bounce off of the ionosphere and that the properties of the ionosphere that control this "bouncing" have a diurnal variation to them. This fact can be backed up nicely with ion density plots obtained from the MIT-Haystack web-site (www.haystack.mit.edu).

To try to get students to understand the fact that non-solid objects can reflect radio waves, I wrap a portable radio in a wire mesh and show how the signals can be blocked.

These activities are all part of a lesson plan I developed at the MIT-Haystack observatory. Details of this lesson plan can be found at the MIT-Haystack web-site at www.haystack.mit.edu. Look in the "Table of Contents" under "Research Experiences for Teachers".

Jeffrey A. Yuhas
Reading Memorial High School

Greek Light Pollution Education Program

From 1997 to 1999 an educational program on light pollution, funded by the Greek Ministry of Education and Religion has been run, The program has met with great success and can serve as a model for other light pollution education programs. The program was coordinated by Dr. Margarita Metaxa, a teacher at the Astrolaboratory Unit of the Arsakeia - Tositseia Schools. As part of the program, an international symposium on light pollution was held in Athens, in May 1999. The proceedings have been published as well as an educational CD-ROM on light pollution. By the end of March 2000, the Light Pollution Bureau of Athens will be fully established and will co-ordinate all the "Light Pollution" activities in Greece.

A major part of the program is the creation of over 20 light pollution study centers at schools around Greece. The Greek program is working with others throughout Europe as well and is anticipating creating an internet-based forum on light pollution.

The program's targets are to familiarize students with:
--Familiarize students with the problems of light pollution
--Study the problems through astronomy, physics and computer science
--Consider the cultural and social dimensions of the impact of Light Pollution
--Appreciate the effects on our heritage and environment throughout our country.

The participating students cooperate with observatories, universities, technical universities, lighting companies, other groups of students, the local authorities, societies and associations in order to study and assess the problem of light pollution in Greece.

For more information about the program, see

http://www.uoi.gr/english/EPL/LP/lp.htm

Stephen M. Pompea
Pompea & Associates

Top 5 Resources on Teaching about Light Pollution

1. Posters (Earth at Night, North America at Night, Europe at Night)
US$10 for each poster, for shipments in the USA; US$15 for shipments outside the USA
Satellite images of the Earth at night create beautiful and educational posters. These maps of the lights as seen from space not only teach geography but also can be integrated into such interdisciplinary curricula as social sciences.

2. Los Angeles Video: Looking to the Stars
US$20 for shipments in the USA; US$25 for shipments outside the USA
The video opens with inspiring astronomical footage and reasons why we should care about light pollution. The City of Los Angeles explains the challenge, efforts, and successes faced by this metropolis with so many roadway miles.

3. All IDA Slide Sets on CD-ROM (JPG, PC Format)
US$15 for shipments in the USA; US$20 for shipments outside the USA
Contains many satellite photographs, examples of good and bad lighting, and text slides. All support presentations on these subjects.

4. Info Sheets on CD-ROM (PC, WordPerfect format)
US$15 for shipments in the USA; US$20 for shipments outside the USA or
Complete Set of Info Sheets
US$30 for shipments in the USA; US$35 for shipments outside the USA
Also available to be downloading from http://www.darksky.org
Activities, background, and fact sheets on many topics (economics, security, sample ordinances, how to talk to the media, etc.) are available.

5. Individual slide sets
Slide Set A with Booklet
US $25 for shipments in the USA; US$30 for shipments outside the USA

IDA's main slide set for general presentations.
Slide Sets B through G (each)
US$35 for shipments in the USA; US$40 for shipments outside the USA
Supplementary slide sets include alternatives to slide set A as well as sets which target specific topics such as good lighting or bad lighting or satellite images.

All can be ordered by contacting IDA at:
International Dark-Sky Association (IDA)
3225 N. First Avenue
Tucson, AZ 85719-2103

(520) 293-3198 (voice)
(520) 293-3192 (fax)
(877) 600-5888 (toll free for membership)
ida@darksky.org (IDA office)

Elizabeth M. Alvarez del Castillo
International Dark-Sky Association

Teaching about Light Pollution

Light Pollution is a growing threat to our nighttime environment, one that has already seriously harmed astronomers, both amateurs and professionals. We are faced with the distinct possibility that in only a generation or two very few people will be able to have a direct view of the universe. Urban sky glow will have blotted out the dark sky, just as a lighted room blots out the view of a slide show.

Components of light pollution include:
1. Urban Sky Glow: It is destroying humanity's view of the universe.
2. Glare: It is blinding us and harming visibility; ***glare is never good!*** It hinders safety and security.
3. Light Trespass: Spill light and obtrusive light occur when light falls where it is not wanted or needed.
4. Clutter: It trashes the nighttime environment and causes confusion as well.
5. Energy Waste: Wasted light costs over One Billion Dollars a year, in the U.S.A. alone.
Quality outdoor lighting uses only the amount of light needed, when and where it is needed.

There are solutions to all of these problems. Quality lighting is the key. These solutions preserve the
dark skies for all of us, improve the quality of the nighttime lighting and the

nighttime environment (better visibility, better safety and security, more attractive surroundings), and save money as well for we use light rather than waste it. We all win!

Defining Good and Bad Lighting

Good lighting means:
No glare! Glare always degrades visibility.
Use the right amount of light, not overkill and not too little.
Use fairly uniform lighting, so our eyes can adjust easily.
Avoid deep shadows. We need smooth transitions from light to dark.
Avoid light trespass. Shine the light only where it's wanted or needed.
Minimize the up light. We don't live up in the sky. Save the stars!
Avoid the clutter and trashy look of areas with poor quality lighting.
Save energy. We waste far too much light; light that produces glare, bright skies, light trespass. We can save over a billion dollars a year in the U.S.A. alone.

How to get good nighttime lighting:

1. Use quality lighting fixtures, ones that control the light output, putting it only where it is needed and wanted. Never use fixtures that spray light everywhere.
2. Use the right amount of light, not overkill.
3. Use energy efficient light sources.
4. Use occupancy sensors, time controls, and dimmers where possible.

Earth at Night Poster

Use our Earth at Night poster to teach geography and social sciences as well as science, technology, and math. It's amazing what these pictures show us about the practices and values of various communities.

Specific Teaching Ideas

1. Energy/Economics estimates
2. Local pollution assessment
3. Surveys with Gratings
4. Surveys of public perceptions
5. Rank areas based on perceived "good/bad" lighting
6. What constitutes good and bad lighting
7. Test listing common misconceptions
    --more light is better
    --security lights prevent crime
    --just leave town to get away from the lights
    etc.

Making a difference on the political arena or in your neighborhood/community

How Faint a Star Can be Seen with the Unaided Eye?
This depends on the ability and experience of the observer, the state of dark adaptation of the observer, the sky brightness due to light pollution from electric lights and natural sources such as the Moon, and the clarity of the sky.

Students can observe several constellations under different lighting conditions. Have them draw pictures in their logs. Another option is to supply a picture of the constellations and have them circle the stars they see. (For ages where competitiveness is a problem, you might insert some non-existent stars so you can tell whether they are only circling those they really see.) For advanced students, this exercise should include presentation of the magnitude scale.

Excellent teacher notes are included in IDA information sheet 120. It includes drawings how three well known constellations would appear for 4 different limiting magnitudes.

Other Projects
1. Write stories and poetry or create an art project showing what the heavens mean to you.
2. Keep a log over the course of a month during which you observe the skies each night. Explain the patterns you see and what variables affect your view of the stars.

References / Resources
The International Dark-Sky Association
3225 N. First Avenue
Tucson, AZ 85719-2103
(520) 293-3198
(520) 293-3192 (fax)
ida@darksky.org
http://www.darksky.org

In addition to many visual education resources, IDA has numerous information sheets on relevant issues.
Specific activities are in information sheets 58, 120, 127, 113. More will be added this year!
Information Sheet 58: Star Watching Program
Information Sheet 120: Light Pollution and Limiting Magnitude
Information Sheet 127: Save the Stars! Activities for Elementary Students, Grades K-3

Information Sheet 113: Count the Stars! An Activity for Elementary Students, Grades 4-6

The Universe in the Classroom, No. 44 - Fourth Quarter 1998
by David Crawford, Margarita Metaxa, and John Percy
The Astronomical Society of the Pacific
This issue is all background and activities on light pollution.

Information Sheet 120: Light Pollution and Limiting Magnitude [Extracted teacher notes]

At an extraordinarily dark site, stars as faint as visual magnitude +7.0 can be seen by experienced observers with good eyes. There are about 14,000 stars brighter than magnitude +7.0, so under ideal conditions an observer could see somewhat under half of them (~7,000) due to only half the sky being visible and atmospheric extinction. A magnitude +6.0 sky is still a reasonably good sky, with ~2,400 stars visible to the unaided eye. There is some light pollution, and it is usually enough to illuminate clouds so that they no longer appear utterly black against the sky as with a magnitude +7.0 sky. The brighter parts of the Milky Way are still readily seen. Even so, only about a third of the stars are visible that can be seen under a magnitude +7.0 sky.

A magnitude +5.0 sky is affected by moderate light pollution, with ~800 stars visible. This is only about a third of the stars that can be seen in a magnitude +6.0 sky, and just 10% of what can be seen under a magnitude +7.0 sky. The Milky Way is barely visible, if at all.

Less than 250 stars are visible in a magnitude +4.0 sky, and the Milky Way is never visible. Light pollution is a serious problem.

A magnitude +3.0 sky will show fewer than 50 stars, and light pollution is severe. This is the typical sky encountered inside a major city.

A magnitude +2.0 sky will show fewer than 25 stars. This is typical of the central regions of the largest cities. On the following three pages, we show you how the appearance of some familiar constellations (Orion, Cygnus, and Ursa Minor) changes as the naked eye limiting magnitude (the faintest magnitude you can see) deteriorates from +7.0 to +2.0 in 1.0 magnitude steps.

Let's take a look at Orion, the Hunter, on page 2. Orion, arguably the most prominent of the constellations, begins to look more like "Orion, the Hunted" under a magnitude +4.0 sky. Under a magnitude +3.0 sky, Orion is on his deathbed. When light pollution is so bad that we have a magnitude +2.0 sky, only blazing Betelgeuse, regal Rigel, and Bellatrix and Alnilam remain to regale us.

Take a look at Cygnus on page 3. Under a magnitude +7.0 sky, our celestial swan is awash in stars, and is flying along the greatest of all rivers - the river of stars we call the Milky Way. Under a magnitude +6.0 sky, the Milky Way in Cygnus is still prominent, but diminished. Lesser-known byways of the Milky Way have disappeared altogether, and much of the intricate structure of the Milky Way is no longer easily discernible. By the time we reach magnitude +5.0, the Milky Way in Cygnus is getting very difficult to see. At magnitude +3.0, the prominent "Northern Cross" asterism of Cygnus is no longer intact: Albireo, perhaps the prettiest double star in the northern skies, is utterly invisible. Under a magnitude +2.0 sky, only the star Deneb remains.

Looking at Ursa Minor on page 4, we see that under a magnitude +7.0 sky, stars can be seen inside the bowl of the Little Dipper. Under a magnitude +5.0 sky, the Little Dipper asterism is getting rather difficult to see. At magnitude +4.0, all that remains of the Little Dipper is Polaris and the two stars on the opposite side of the bowl, Kochab and Pherkad. At magnitude +2.0, even Polaris is extinguished, and we are left looking for darker pastures.

Information Sheet 58: Star Watching Program

A detailed chart of the Pleiades star cluster is included in this information sheet. Only the brightest (biggest) stars on the chart will be visible to the naked eye.

Find the Pleiades Cluster in the sky; use any star chart if you can't find it from your own experience.

If the sky is bright with urban sky glow, you may well have a difficult time of it. Any monthly issue of Sky & Telescope or Astronomy magazine has an excellent monthly star chart (you need a winter month), as do many books on astronomy. Check out your library or bookstore if you don't already have a chart that shows the Pleiades on it.

Remember that the program is not a test of eyesight, but of sky conditions. We need data from the center of the largest cities, suburbs, smaller cities, as well as from excellent dark sky sites in the country. Data from the same observer on different nights and from different sites are most welcome.
We appreciate very much your input.

Visual estimates should be made with both the naked eye and with binoculars (if you have some) on
clear moonless nights when the Pleiades are rather high in the sky. Wait some

minutes before beginning so your eyes can dark adapt. Circle the stars on the chart that you can see with your naked eye; put a small cross next to the ones you can see with your binoculars. Enter your observing data on the form given below.

Feel free to copy and use this form as much as you like. Distribute it to others as well.

Data to Accompany the Pleiades Chart

Date: ________________ Local Time: __________________ to ____________

Name:

____

Occupation:

____

Age: ____________ Sex: (M / F): ______ Eyesight: ____________________

Address:

____

____

____

Place of observation (be as specific as possible):

____

____

Comments on the location (Example: Only one street light within 200 yards and it is well shielded from my view):

____

____

Comments on sky condition (Example: Light clouds to the north, some haze, but about normal for this location):

____

____

Comments on the observer:

1. Binocular specification (Example: 8 x 50, or 7 x 50, or 10 x 70):

Spec: __________ Make of binoculars: ________________

2. Notes (Example: Used my regular eyeglasses, my vision is about 20/20. Used averted vision for the naked eye estimates.):

____

____

The Star Watching Program is described and discussed in the November 1991 issue of *Sky & Telescope*. In the November 1992 issue is a discussion of a similar Star Watching program that has been running in Japan for several years now and upon which the IDA program is based.

The International Dark-Sky Association has received hundreds of responses in the years that the program has operated; while in Japan, the responses have been in the thousands. We hope that in future years the program will grow extensively in the U.S.A. and in other countries. We appreciate your interest and support! Dark skies are wonderful!!

Elizabeth M. Alvarez del Castillo
International Dark-Sky Association (IDA)

Chapter 11
The Sun

Sunspots

As one of our undergraduate astronomy laboratories, we have the students determine the rotational period of the sun by observing the apparent daily motion of sunspots across the solar disk (seen in projection through a telescope onto a piece of paper). Since they see the spots on a two-dimensional surface, and the spots are actually on a three-dimensional surface, the students must make a construction of where the spots would be on the surface of the sun, ie., they make a "pop-up" diagram of the sun at the latitude at which the sunspots occur.

Since some students find this concept difficult to understand, I use an orange to illustrate the idea: I cut a quarter wedge out of the orange, and show the orange to the students while keeping one of the cuts horizontal. I then draw the position of a sunspot on the orange at the horizontal cut, and project the spot back to the inside of the orange on the vertical cut, to show where the spot is when projected onto a two-dimensional surface.

Joanne Rosvick
University of Victoria

The Luminosity of the Sun

It is instructive to show the students a simple way in which they can determine the luminosity (power) of the Sun. Bring a light bulb to class, mounted on a piece of wood (or, just bright a lamp without the lamp shade). For simplicity, have this be a 100-watt bulb --- i.e., 10^9 ergs/s. Ask a student to come to the front of the room and place his/her hand close to the bulb, adjusting the distance until the hand feels as warm as when it is exposed to direct sunlight on a bright, sunny day. Estimate the distance between the hand and the bulb. Now take the ratio of the Earth-Sun distance to the bulb-hand distance, square it (because of the inverse-square law of light), and multiply by the power of the bulb (100 watts). The result is the power of the Sun! For example, suppose that the hand-bulb distance is estimated to be 7 cm. Then L(Sun)

$= L(\text{bulb}) \times [(d_{\text{Sun-Earth}})/(d_{\text{bulb-hand}})]^2$
$= (100 \text{ watts}) \times [(1.5 \times 10^{13} \text{ cm})/(7 \text{ cm})]^2$
$= (100 \text{ watts}) \times (4.6 \times 10^{24})$
$= 4.6 \times 10^{26}$ watts
$= 4.6 \times 10^{33}$ ergs/s, comparable to the true luminosity of 3.9×10^{33} ergs/s.

Alex Filippenko
University of California, Berkeley

Diameter of the Sun

Many important aspects of the scientific method can be illustrated by a project to measure the diameter of the Sun with a pinhole camera. The size of the image divided by the distance between screen and pinhole gives the same ratio as the diameter of the Sun divided by the distance between the Earth and Sun (AU = 1.496 times 10^8 km). Simply multiply the ratio you measure by the value of the AU to get the diameter of the Sun. Of course, the experiment is a bit of a con since the ratio of the Earth/Sun distance must be given in advance, but this does not lessen its effectiveness for the students. Instruct them to make a pinhole in a thin piece of cardboard, with the usual dire warnings about not looking at the Sun directly (to save eyes and ward off lawsuits).

Tell them to use another white cardboard as a viewing screen, and hold the pinhole above the viewing screen to form an image of the Sun. They should measure the size of the Sun's image and the distance between the pinhole and the viewing screen (this is easier with two people), measure several times, record the answers, and take an average value to improve accuracy. How does their answer compare with the published value? How could they improve the accuracy of this measurement? Can they see any sunspots in your image? Without leading them too much, students will often come up with astute insights into the nature of systematic and random errors, and simple principles of optics.

Chris Impey
University of Arizona

The Setting Sun

Most people now live in urban areas, and have only the dimmest awareness of the fundamental motions of the sky. A simple but effective exercise (for extra credit or as a project) is to get students to track the location of the setting Sun over a period of five weeks (few students get up early enough to observe the rising Sun!). Tell them to choose a location where they have a reasonably clear view of the western horizon, and where they can return to exactly the same place later. Instruct them to make a careful and neat sketch of the horizon detail to scale, and to mark the exact position of the Sun at sunset (reminding them of course to be very careful never to stare directly at the Sun!). They should repeat this observation once a week for five weeks, and write a report to describe and explain their observations. Remarkably, in any large introductory class, you are likely to find a few students who insist that the setting position did not move. As mundane as it is, many students are deeply impressed by this exercise.

Chris Impey
University of Arizona

The Photosphere of the Sun

A difficult concept is the idea of a photosphere. Make an analogy with clouds and fog to show that there can be a surface of last scattering of light that need not correspond to a large change in density. From center of sun to surface, photon takes a million years, degrading from X-ray to optical wavelengths, then eight minutes to us at the speed of light. Point out that despite the overall constancy of the Sun's power, its subtle changes have dramatic consequences on Earth (radio interference at sunspot maxima, droughts at sunspot minima).

Chris Impey
University of Arizona

What Is a Random Walk?

Most electromagnetic energy gets out of the sun in a very roundabout way. The photons that we see as sunlight actually had a heck of a time leaving the interior of the sun. They did what physicists call a random walk. It's kind of like trying to make your way through a crowded subway station. You head off in one direction, get bounced in another direction, then still another, and so on. The problem is that, unlike the person in a subway station, the photons (particles of light) don't know which way they want to go. They are jostled about in scattering collisions with particles (mostly electrons). It's now more like being blindfolded in a crowded subway station. Eventually you will work your way to the door by chance and leave. But how many steps will it take you to go a certain distance?

A demonstration that can be used in class illustrates this. One person acts as the photon and we see how many steps it takes to travel ten paces, if the person is moving in a random direction on each step. The instructor draws a circle around the student ten paces in radius and now takes bets from the class. You know it will take a minimum of ten steps. We will use a little compass wheel to tell him/her in direction to take each step.

It turns out that to go ten paces will require $(10)^2$ or about a hundred paces, if each pace is in a random direction. Restated differently, a random walk of N steps, on the average, will move you a distance of $N^{1/2}$ steps away. So, it should take about 100 steps to get across a circle ten steps in radius.

For the sun, radiation leaves the center, where it was generated, by a random walk process. A photon moves about 1 cm before it has a collision and is randomly redirected. Since the radius of the sun is 10^{11} cm, a photon must travel about 10^{22} cm before it will make its way to the surface. Since the photon travels at the speed of light (3×10^{10} cm/sec), it will take about 10^4 years for a photon to emerge. A much more detailed calculation gives a time of 10^7 years, since the mean free path of a photon is not always one centimeter throughout the sun.

This exercise implies that changes deep in the sun will take a considerable amount of time to be reflected in the light that we see. This is true. Neutrinos, which are not scattered in the same way as photons (neutrinos go right through matter without interacting), pass directly out of the sun at the speed of light. They reflect very rapidly any changes in the interior of the sun. They are also very hard to detect, since nothing wants to absorb them. A great many neutrinos pass through our body every second without any interaction or effect. But that's another story.

Stephen M. Pompea
Pompea & Associates

Why Sunspots Are Darker (Because They Are Cooler)

The appearance of dark sunspots on the sun surface is explained by the fact that they are cooler than the background areas and thus appear darker because they do not radiate as much power at the lower temperature. This effect can also be shown using two light bulbs. One of lower power as 7.5 watts will actually appear darker than one of higher power as 75 to 100 watts when held up in front of the brighter one. The difference is not as noticeable when both bulbs are placed at the end of the lecture table, which I do before bringing them together.

James Kettler
Ohio University - Eastern Campus

Solar Luminosity

There are three parts to this activity. During the first part, students use a wax cube and aluminum foil as a simple photometer and determine the solar luminosity by comparing the power output of the Sun with the known power output of a light bulb. During subsequent sessions, students record how quickly the Sun heats up a container of water in order to calculate later the solar constant, a measure of the amount of solar radiant energy at the top of our atmosphere. The experiment here is simple: students sit outside for about an hour monitoring the rise in temperature of a cup of water aimed at the Sun. However, you may find that the scientific analysis of these measurements is a bit more sophisticated. With the help from Einstein's equation $E=mc^2$, students will project their value of the solar constant into the Sun's inaccessible interior to calculate the rate at which mass is being consumed in nuclear reactions to sustain the Sun's radiant energy output.

Constance Walker
University of Arizona

Chapter 12 Properties of Stars, Star Clusters, and the Interstellar Medium

Stars and Balloons

Stars are made out of gas, and they are held together by the force of gravity. Those that are stable have found a balance between the force of gravity, trying to pull them together, and the gas pressure, trying to expand them.

Balloons are made out of gas, and they are held together by the elastic force of the stretched rubber. Those that are stable have found a balance between the force of the stretched rubber, trying to pull them together, and the gas pressure, trying to expand them. It is no accident that the two paragraphs above are nearly the same: the processes of finding the conditions of force balance are the same in both star and balloon.

A star is hot, and so it radiates energy into space. This causes it to lose heat energy, which is where the radiation energy comes from. When a normal gas loses heat energy, it also loses some of its ability to exert pressure, holding it up against the force of gravity; thus it cannot maintain force balance, so it must contract to a smaller size. It is like a leaky balloon: the air leak of the balloon has the effect of the radiation leak of the star: star and balloon maintain their gas pressures against their leaks.

Suppose that we pump air into the balloon at just the correct rate to make up for the air lost through the leak. Then it can maintain the pressure needed to keep it in force balance, and it will not shrink. This is exactly what the nuclear reactions in a star do: they release nuclear energy that is quickly changed into heat energy. The star must adjust itself so that the nuclear energy release exactly balances the radiation leak at the surface, i.e., it must come into energy balance. Then it can keep the pressure needed to stay in force balance, just like a leaky balloon connected to an air pump.

A star that appears to be simply sitting there doing nothing is actually in a complicated balance between many competing physical processes.

Thomas L. Swihart
University of Arizona

A Light Bulb "Shell Game"

The distinction between apparent magnitude and absolute magnitude (brightness and luminosity) can be illustrated before a class with three or four common light bulbs of differing wattages. I place each bulb in turn into a live socket at the end of a long extension cord so that the class can see how bright the bulbs appear from the front of the classroom. I then ask a student to play a guessing game with me: Can she or he name the wattage when a random bulb from the set is placed into the socket? The game is easy, as long as the socket remains in one place and the bulbs are all observed at the same distance.

After one or two "winners" I change the game on the next unsuspecting student. I run around the classroom carrying the bulb and socket! By constantly changing the distance between light source and observer, the distance becomes ambiguous. All the while, of course, I encourage the "contestant" not to hesitate and to come up with the wattage; after all, it was so easy for the previous two! He/she will usually still arrive at the right answer, but not until I have brought the bulb back to its original distance from the student. The moral? When distance cannot be used, absolute magnitude/luminosity becomes difficult to predict.

Finally, I ask students to image a "light bulb" that appears as a mere point (so the apparent size of the bulb itself cannot be used to judge distance) and one in which the possible ranges of wattage and distance are orders of magnitude greater than that which I can duplicate in the classroom -- the very problem faced by astronomers when studying the stars. A warning: the most difficult part of this demonstration, which works best in a darkened classroom, is screwing into the socket the light bulb and not the instructor's finger!

Thomas Hockey
University of Northern Iowa

Observing the Color of Stars and Comparing Them

Using three or four illuminated light bulbs of equal wattage but of different colors, all placed on a desk, let the students look at the bulbs through colored transparencies. The blue light bulb will appear brightest when viewed with a blue transparency and the red light bulb with the red transparency, etc. After this demonstration in the classroom, the students can then look at the stars at night through the colored transparencies and thus pick out the blue, red, yellow stars, etc. They can then be taught the relationship between color and temperature.

A better result is obtained if a Polaroid camera is used to photograph various first and second magnitude stars with first a blue and then a red filter (the same colored transparencies can be used that were used in the visual sightings and in the classroom) for equal exposure time, and the photographs are compared. If both a blue and red star are on the same photograph, one easily observes the difference.

V. Gordon Lind
Utah State University

Illustration of the Forces and Mechanisms of Fusion

This is a little demonstration to illustrate how the Coulomb barrier is overcome in the proton-proton reaction in the interior of stars or any other fusion reaction. Use two cylindrical magnets of about 6-8 inches long each. Place Velcro around their ends. Place one magnet at rest and roll the other towards it. Arrange the orientation of the magnets so that the like poles will repel and the two magnets roll away from each other. If the velocities with which they are rolled together is high enough, the magnets will contact each other through the Velcro tips and stick together.

The magnetic repulsion simulates the Coulomb repulsion between the two protons, whereas the sticking of the Velcro simulates the strong nuclear force that binds the two protons together. Unfortunately, the weak nuclear force is not illustrated by this simple demonstration. It enables one of the protons to disintegrate into a neutron, positron and neutrino while the two protons are in contact and then the binding between the neutron produced and the remaining proton becomes permanent as a deuteron is formed.

In order to give some quantitative control on the velocity with which the two magnets come together, use a narrow inclined plane for the one magnet to roll down while the other magnet is at a fixed position at the bottom. The tilt of the incline and the distance to travel can be changed to give the collision characteristics desired.

V. Gordon Lind
Utah State University

Star Clusters and Their Evolution

To give students an idea of how unchanging the universe in general is in one person's lifetime, I use this analogy: a mayfly lives for only a day. If the fly is born in the morning, and sees a person, that person has a certain appearance, which is noted by the fly. When the fly nears the end of its life toward the evening of the same day, and sees the person again, the person's appearance has not changed at all, according to the fly.

The fly has lived its whole life and not seen its "universe" change at all. So it is with us and the stars: if we were to observe a cluster and construct an HR diagram when we were infants, say, and then observe the same cluster with the same accuracy and plot another HR diagram when we were old and gray, on the whole we would see no change in the diagram at all, and therefore could not conclude anything about how the stars are evolving. We must observe lots of different clusters to get an idea of how stars and clusters evolve, just as the mayfly must look at lots of people of different ages to see how humans change.

Joanne Rosvick
University of Victoria, Victoria, BC, Canada

Alcohol in Space

Many remarkable molecules have been discovered in interstellar space, including the amino acid glycine (NH_2CH_2COOH), which is one of the basic building blocks of life. In the mid 1970s, astronomers detected large amounts of ethyl alcohol near the center of our galaxy. Ben Zuckerman at UCLA pointed out that the Sagittarius B2 molecular cloud, if purged of impurities, would yield approximately 10^28 fifths at 200 proof, which exceeds the sum total of all human fermentation efforts in recorded history. A few years after this discovery, NASA received a proposal from a Texas businessman to "harvest" this alcohol. The diffuseness of interstellar space makes any such proposal impractical. It turns out that a spaceship pushing a 1-km diameter funnel at 10% of the speed of light would take 1000 years to collect enough ethyl alcohol for one martini!

Chris Impey
University of Arizona

H-R Diagram

The Hertzsprung-Russell diagram can be better understood when an analogous diagram is drawn. For example, a similar diagram of human attributes can be drawn which in no way implies motion through space of time.

Prepare a diagram of heights versus weights for a group of people. There will be a "main sequence" where most individuals lie. Those above the line might be classified as giants; those below it as dwarfs. Though we cannot see these individuals directly, we can deduce their characteristics from other knowns. For instance, even though we can't generally directly measure a star's diameter, it can be inferred from our knowledge of the Stefan-Boltzmann law.

Carl J. Wenning
Illinois State University

H-R Diagram

The H-R diagram is so powerful and contains so much information that sometimes a diversion on the use of graphs is necessary. Hypothetically, plot happiness versus wealth for a population of people, point out the existence of a correlation and the likelihood of points elsewhere in the plot (i.e., poor but happy, rich but miserable). Importance of H-R diagram is that points do not lie just anywhere on it. Use two examples to show that two quantities can be related only through a third. First, plot the relation between children's height and their IQ, where the third parameter is age (are taller children smarter? no, just older). Second, plot a correlation between the size of feet in China and the price of fish in Billingsgate market in London, where the third parameter is just time.

Then return to the first example with a plot of age versus wealth for a population. In this analog H-R diagram, there is a main sequence, corresponding to the median growth of income with age, and other clumps of haves and have-nots. Note the form of such a diagram is peculiar to a certain population (it would look very different for the US. and India and Saudi Arabia).

Chris Impey
University of Arizona

H-R Diagram Tip

The H-R diagram is often confusing to students the first few times they see it. They have trouble keeping track of what is being plotted against what. Students in the back of a large lecture room might have trouble reading the axis labels from an overhead transparency.

A small investment in labor will pay off many times to avoid any confusion in the H-R diagram. Color a transparency (or buy some strips of colored acetate) so that the colors of the rainbow are illustrated on a single sheet. (Or make a single color photocopy of the colors of the rainbow.) Overlay this sheet with every H-R diagram you use on the overhead projector, and there will never be any confusion as to what is being plotted on the horizontal axis. The stars in the red giant region will be in the "red" portion of the H-R diagram.

Stephen M. Pompea
Pompea & Associates

Population of Stars

Two concepts confuse students when discussing population of stars. One is that while dwarfs are not common in a volume-limited sample, main sequence stars and giants are more common in a flux-limited sample. This can be demonstrated schematically with a drawing of 50 and 100 watt light bulbs distributed at different distances from the observers. Or comment that in any crowd, we notice the few loud voices more clearly than the silent majority. The second is distinguishing the evolutionary phases of stars. Point out that we see evolutionary phases in proportion to how long they last; the brief moments of star birth and death are rarely seen. These two factors can be combined in the statement that for any stellar population, the visibility of a phase of evolution is proportional to the length of the phase times the luminosity of the phase.

Chris Impey
University of Arizona

Between Stars

Point out the extreme diffuseness of space. For example, from sea-level air to the best lab vacuum to the space between galaxies, the density of particles goes from 10^{25} to 10^{18} to 10^{9} to 0.1 per meter^{3}. The corresponding distance between particles goes from 10^{-9} to 10^{-6} to 0.01 to 2 meters. Nevertheless, the enormous path length gives big effect from the scattering and reddening of light. This leads into discussing dust and its effect on distance measurement, and scattering and its effect on everyday things like sky color, sunset color, etc.

Chris Impey
University of Arizona

Useful Demonstrations

A colorful demonstration of "nebulae" is done by using a medicine dropper to add drops of food coloring (red, blue, green) to water in a petri dish or any shallow transparent container placed on the overhead projector. Heat from the overhead precedes motion of the "gases" of the nebula.
Magnetic fields in space are demonstrated on the overhead with iron filings and a magnet placed under a piece of Plexiglas on the overhead.

Gene Maynard
Radford University

Distance Scale

As an analogy of the overlapping chain of distance indicators used in the universe, consider how we would estimate distances standing on top of a building. To the edge of the roof, a meter stick gives an accurate reading. To the nearest few hundred yards we use our estimate of the size of people. Up to a few miles away, the apparent size of buildings gives a clue. Finally, the appearance of the distant horizon gives a crude guess--as we can't see individual features (trees) at the horizon, we can't see individual stars in distant galaxies--as smog and dust in the air affects our estimate of the distance of the horizon, so dust in the universe affects our largest distance estimates. This analogy should clearly illustrate the accumulating errors and assumptions involved.

Chris Impey
University of Arizona

Spectral Classification

A useful non-computer demonstration or class exercise is to have students classify stellar spectra. Rather than forcing students to a priori fit their spectra into rigid classification schemes, I tell the students to merely divide the spectra into meaningful groups. Most students will divide the 12 or so spectra into 3 or 4 groups, with little substantial variation between students. I then introduce the MKK system and have the students reclassify the spectra. Commonly, students need to make only slight adjustments to get their spectral sequence to follow the MKK scheme.

Another favorite class exercise is to have students use actual stellar flux curves when I talk about blackbodies. I use either Anita Cochran's monograph, *A Catalogue of Energy Distributions*, or the University of Washington's, *A Display Atlas of Stellar Spectra*, as the source for the blackbody curves. From the flux curves, I have the students estimate the star's temperature using Wien's law (and using Planck's law in advanced classes). The students get a feeling for working with real data and that stars are not true blackbodies.

David Bruning
University of Louisville

Apparent and Absolute Magnitude

The differences in the apparent magnitude and absolute magnitude of a star is a difficult concept to comprehend. An idea that can be used in the classroom to demonstrate the concept is the following.

Take two light bulbs of visually different brightness (as a 7.5 watt and a 60 watt). Connect each to a socket with a long extension cord. Locate the brighter bulb far from the observer and the dimmer one closer so that the dimmer appears to be brighter than the bright bulb. Thus the apparent magnitude as seen from earth! Now locate the bulbs at a standard distance from the observer and see how they "really" compare. The brighter one is "really" brighter. Thus the absolute magnitude is demonstrated. One can also use a photometer or photocell connected to a milliammeter to measure the light output of either bulb as the distance from the detector is doubled, tripled, etc. For variety, one can use colored bulbs--red, orange, blue--to approximate the color of stars.

James Kettler
Ohio University - Eastern Campus

A Proper Motion Demonstration of a Dwarf Star with a Dark Companion

For this device I use a portion of an old sponge-mop handle. The sponge holder is removed, leaving a metal collar into which one can easily insert a standard piece of chalk. The handle is cut about 18 inches above the collar. On the collar end of the handle I thumbtack a circular cardboard disk (about 7" in diameter) colored red (to represent a red dwarf star) on the side of the handle opposite the chalk. On the other end I thumbtack a black cardboard disk (about 2.5" in diameter) to represent a dark companion. Finally, a wood screw with a loose metal collar is inserted into the handle about 1/3 of the distance from the red disk to the black disk, on the disk side of the handle.

For the demonstration I draw a straight line across the blackboard representing the motion of the center-of-mass of the system (to be represented by the wood screw position). I then grab the (raised) thumbtack, holding the red disk in one hand, and holding the wood screw collar pivot in the other hand I move the device along the straight line with the pivot point following the straight line, and the other hand rotating the red disk about the black. With the chalk in contact with the board, the sine-like wave motion of the red dwarf through space is demonstrated. The black companion could be construed as either a massive planet or a brown dwarf, causing the oscillatory motion of the red dwarf through space.

Richard D. Schwartz
University of Missouri - St. Louis

An Astronomical Measurement Technique - Parallax

Anyone raised on a farm (or who has kept a pet chicken) will remember the peculiar leisurely walk of a chicken: the chicken's head remains stationary when the body moves forward, then the head moves forward with the body at rest, then the head stays fixed again while the body moves on, and so on. The "chicken walk" can be mastered with a little practice and makes an amusing demonstration of an astronomical measurement technique.

The chicken has another peculiarity related to its walk that anyone who has held a calm chicken will remember. If the body is moved by a small amount in any direction, the head will not follow immediately, but will remain fixed as if anchored in space. If the body is moved around in a horizontal plane, the head will stay in the same spot as if completely isolated from the body by a frictionless neck.

Why does the chicken have this ability, and why does it walk that way? I explain to my astronomy classes that it is because the chicken has eyes at the sides of its head. Therefore, the chicken has only monocular vision of its surrounding, like the astronomer's view of the stars. Yet the chicken has a need to know distances to objects (like worms), just as the astronomer needs to know distances to stars.

The chicken measures distances during its walk by using the shift in positions of images on its retina when the head moves through a standard distance (one step). Astronomers measure distances to stars by noting shifts of images of stars on photographic plates when the telescope moves through one Earth orbit around the sun. Both are using parallax to measure distance.

While the chicken's body moves through one measured step of the walk, the neck flexes automatically to keep the images on each retina fixed in place; the change in position of an image on the chicken's retina during the following rapid forward motion of the head allows an estimate of the distance to the object. Replace the chicken step by an Earth orbit and the chicken's eye by a telescope and the analogy is complete.

Ronald Stoner
Bowling Green State University

Chapter 13
Stellar Evolution

Black Hole Model

Two students hold the corners of a sheet of latex rubber that has dimensions of a least 100 cm x 100 cm. The sheet is held so that it is horizontal and the instructor then drops in a heavy steel ball bearing or a small lead sphere. The resulting distortion of the sheet of rubber demonstrates how the shape of space is distorted by the gravitational field surrounding a black hole.

Gene Maynard
Radford University

White Dwarfs, Neutron Stars, and Black Holes

It is usually explained in class that when a star reaches the end of its lifetime the ultimate fate which befalls it depends upon its mass. White dwarfs are always less than about 1.4 solar masses. A stellar corpse of mass greater than this but less than about 3 solar masses will become a neutron star. Even heavier stars must end their lives as black holes. However, the physical size of such objects as described are related inversely to their mass. Students have difficulty grasping how a more massive object can be physically smaller than a less massive one. The key of course is density.

To show the comparison of these objects of typical characteristics I use three spheres: One is of Styrofoam or foam rubber and is about the size of a similar earth globe which I use to show comparison of the white dwarf to the earth. Another sphere represents the neutron star and is a wooden or aluminum ball about 1 inch in diameter. To represent the stellar black hole a small steel bearing or lead ball is used. The exact dimensions are not as important as the fact the larger ball has less weight than the smaller ones, with the smallest one the heaviest of all three. By squeezing more foam into a smaller size package one can even show that white dwarfs with larger mass have a smaller radius.

James E. Kettler
Ohio University - Eastern Campus

Simulated Pulsars

Pulsars are neutron stars that are oriented in such a way that their rapid rotation produces a lighthouse beam of electromagnetic radiation pointed in our direction. This radiation as received on earth will pulse periodically in tune with the rotation rate and is very regular. To simulate this in the classroom, one can use a photocell connected to an oscilloscope or computer having suitable software and printout capabilities. Point the solar cell toward the overhead fluorescent lights and the 60-cycle voltage supply will provide a regular flicker of light intensity that can be picked up and displayed. Using the time base on the oscilloscope, the time interval can even be determined and found to be in the millisecond range--a millisecond pulsar ! The computer output can also be analyzed accordingly. Alternatively, one can use the flasher modules used on Christmas tree lights and detect the pulses of light visually as well as electronically.

James E. Kettler
Ohio University - Eastern Campus

A Demonstration of the H-R Diagram

On a sheet of graph paper plot up the color and magnitude of as many of the visually brightest stars that you have patience for. A good source for this listing is the *The Observer's Handbook* of the Royal Astronomical Society of Canada. You can point out to the students that the resulting scatter plot looks nothing like the textbook H-R diagram. If your class is clever you might ask them to figure out why this ought be so. (Hint: it's the distances.) Next plot the same stars, but this time use their absolute, and not apparent magnitudes. The difference will be striking, and can be used to explain why the study of clusters (with stars all at the same distance) is so important to the understanding of stellar properties.

William G. Weller
Association of Universities for Research in Astronomy

HR Diagram - Stellar Evolution

When we trace out the evolutionary path of a star on the HR diagram, many students have the misconception that this somehow represents a motion of the star through space. Point out that it merely represents a change in the physical properties of the star as it evolves. A good analogy to the HR diagram that seems to help my students is a height-weight diagram for people. Many textbooks or study guides include one in the section on HR diagrams. If yours doesn't, make your own. Have each member of the class turn in a sheet of paper with his or her height and weight (include sex if you want a population I and II analogy). If the class is small, one can make the diagram in class; otherwise, it must be made up ahead of time on a transparency. Point out the "main sequence" of early adulthood. Also point out how we would evolve along this height-weight diagram during various stages of our lives: infancy, childhood, adulthood, middle-age bulge, pregnancy, etc. This process demystifies the HR diagram and evolutionary paths.

Also be sure to point out when discussing stellar evolution, that astronomers use the term evolution differently than biologists. The biological context is likely to be more familiar to most students and could easily lead to a misconception about stellar evolution. If we used the term evolution in the same way biologists do, stellar evolution would refer to the changes that take place between generations of stars rather than the changes within an individual star. I find that this idea needs to be explicitly pointed out to many students.

Paul Heckert
Western Carolina University

Neutrino Burst

A remarkable feature of a supernova explosion is the neutrino burst. Neutrinos are the weakly interacting, massless particles that are produced in nuclear reactions of many kinds, including those in the Sun's core. During the core collapse, protons and electrons are forced to merge, creating a sea of neutrons and a huge number of neutrinos. The neutrinos flood into space at the speed of light, carrying away a huge amount of energy, about 10^47 Watts, equivalent to the mass-energy of many Earths. For a brief few seconds, a supernova exceeds the luminous intensity of the entire rest of the universe! The first trace of the recent SN 1987A in the Large Magellanic Cloud came from the neutrino burst. The unseen wave contained so many neutrinos that about 10 billion passed through the body of every person on Earth (in the United States these neutrinos passed up through our feet, since the Large Magellanic Cloud is in the southern hemisphere). As weakly as neutrinos interact, it is an interesting fact that if you were within 40 A.U. of a supernova, neutrinos would kill you by massively disrupting your cell structure (even before the blast wave reached you!). At the distance of the Large Magellanic Cloud, only 1 in 10,000 people suffered even a single (harmless) neutrino interaction. It is even possible that some people suffered a neutrino interaction in their eyeball, releasing a burst of Čerenkov radiation and allowing them to momentarily "see" the supernova! Sensitive detectors in Japan and Ohio detected a total of 19 neutrinos. The detection of neutrinos marked the first experimental confirmation of a 50 year old theoretical prediction -- and the birth of the field of neutrino astronomy.

Chris Impey
University of Arizona

Hydrostatic Equilibrium

As an analogy for hydrostatic equilibrium, bring in a balloon and a large flask of liquid nitrogen. Immersed in the liquid nitrogen, the half-filled balloon will shrink to a new smaller equilibrium state, and it can be pointed out that the compressive force of the rubber (analogous to gravity) has been matched by the smaller amount of pressure from the kinetic energy of gas molecules in the balloon. The other useful analogy is a bicycle pump. Leaning on it with the valve covered up will cause the barrel to become warm as the air column is compressed (some students may not have ever noticed that the column cannot be compressed to nothing).

Chris Impey
University of Arizona

A Common Misconception about Black Holes

Common misconception is that black holes will devour matter voraciously. Point out that gravity law far from a black hole is still Newtonian, that a 1 solar mass black hole of course has the same mass as a 1 solar mass star, so that if the Sun were instantaneously replaced with a black hole of the same mass, the Earth would not waver in its orbit. Point out also that the cross section of a collapsed object to collisions is very small, so only close encounters lead to dramatic effects.

Chris Impey
University of Arizona

The Explosion Mechanism of a Type II Supernova

A Type II (hydrogen-rich) supernova is a massive, highly evolved star whose iron core collapses to form a neutron star. The explosion is powered at least partly by a shock wave formed when the layers immediately surrounding the neutron star bounce off its surface. The neutron star is not yet in an equilibrium configuration; it is rebounding after initial collapse because supernuclear densities were reached. Some energy from the rebound gets transferred to the surrounding layers, thereby expelling them. This is the so-called "prompt" mechanism. (It is now thought that neutrinos, which are released in vast quantities, also help produce a successful explosion after a short pause.)

A great way to provide a heuristic understanding of the non-intuitive "prompt" explosion mechanism is as follows. Get a basketball and a tennis ball. Drop each of them separately, showing that they don't bounce any higher than the height from which they were dropped. This agrees with our intuition. Explain that the basketball bouncing off the floor is analogous to the neutron star rebounding from itself, and that the tennis ball represents the surrounding layers of gas. Now drop the tennis ball and the basketball simultaneously, with the tennis ball on top of the basketball and in contact with it. The tennis ball will bounce astonishingly high. This is a real crowd-pleaser, and it nicely illustrates the prompt explosion mechanism. (I learned this trick from Stan Woosley of UC Santa Cruz. I'm not sure where he learned it. However, I know that it has been described in *Scientific American*, although supernovae are not specifically mentioned.)

For an advanced class, you can expand the discussion. The tennis ball, of course, is very small, yet a massive star has much gas outside the inner core. Thus, the prompt mechanism alone probably does not succeed most of the time. Neutrinos are likely to come to the rescue. After all, they carry away 99% of the energy of the explosion, so only 1% of their energy has to be transferred to the ejecta in order to produce the observed kinetic energy of the ejecta. It seems possible to do this, although the details are still being worked out.

Alex Filippenko
University of California, Berkeley

Some Neat Facts about Type II Supernovae

An interesting fact is that only about 0.01% of the energy of a Type II supernova comes out in visible light. Thus, although for a few weeks a supernova can rival the visual brightness of its host galaxy, the optical display is only a sideshow to what is really going on: the release of tremendous numbers of neutrinos.

Another fascinating tidbit is that during core collapse, which we will assume lasts one second (it actually lasts even less than this!), the amount of energy liberated by a Type II supernova is roughly comparable to that of all normal stars in all galaxies in the observable Universe during that one second. Wow! (The energy is equivalent to about 0.1 solar masses of material, using Einstein's famous equation $E = mc^2$; it is the binding energy of a 1.4 solar mass neutron star.) This is even more sobering when one realizes that core collapse occurs somewhere in the observable Universe roughly once per second. Hence, averaged over time, core-collapse supernovae liberate about as much energy as all the normal stars in the Universe, although primarily in the form of neutrinos rather than visible light.

Alex Filippenko
University of California, Berkeley

Thermonuclear Runaway: Novae and Type Ia Supernovae

White dwarfs in binary systems undergoing mass transfer can behave in several violent ways. Two of the most interesting phenomena are novae and Type Ia supernovae. In a nova, material stolen from the companion star collects on the surface of the white dwarf, and eventually undergoes degenerate thermonuclear burning. In a Type Ia supernova, the white dwarf mass slowly grows to the Chandrasekhar limit, at which point a thermonuclear runaway begins near the center of the degenerate star and literally obliterates it. Radioactive nickel (along with other elements) is synthesized during the supernova explosion, and the decay of the nickel to cobalt and finally to iron produces most of the visible light during the subsequent decline.

An effective way to demonstrate thermonuclear runaway is the trick that is often used in physics classes to clarify the nature of fission bombs and chain reactions. Set up a bunch of densely-packed mouse traps inside a clear Plexiglass box having a hole at the top. There should be two rubber stoppers (corks) on each mouse trap; these will be ejected when the trap is released. Drop a single rubber stopper through the hole in the closed box; a chain reaction is initiated and almost all of the traps are released in a very short time. The display is quite startling.

Alex Filippenko
University of California, Berkeley

Millisecond Pulsars

Quite a few millisecond pulsars are now known. For example, PSR 1937+21 as a period of 1.56 ms (frequency = 642 Hz), and PSR 1953+29 has a period of 6.1 ms (frequency = 163 Hz). The former corresponds to E-flat in the treble clef, and the latter corresponds to E in the base clef. You can bring the appropriate tuning forks to class, giving students an idea of what these pulsars would "sound" like if their spin rate were transformed to a vibrating membrane or fork. With enough pulsars, one can construct a millisecond pulsar "symphony"! (I think something like this has already been attempted by pulsar researchers.)

Alex Filippenko
University of California, Berkeley

Analogy for the Ejection of Matter from the Black Hole

A Garbage Disposal: Remind the class what happens when they put a piece of melon rind in the "garbarator" without putting the lid on. Most goes into the drain, but small pieces come flying out much faster than they went in. So too does a small portion of matter come flying out of the vicinity of a black hole as most of the material falls in.

Firecracker Analogy of the Helium Flash: Remind the class of two burning situations with a firecracker. In one, the firecracker is intact and when lit, it explodes: "KABOOM." In the other, the paper is cut and unrolled and then, when the firecracker is lit, it just goes "fushusshuss" for a while. So it is with helium; when it is compressed to the state of degeneracy, it burns all at once like an intact firecracker; but in more massive stars when the ignition temperature is reached before much compression has occurred, it burns more slowly (actually being self-regulated by expansion of the helium gas as the burning occurs.)

John Broderick
Virginia Polytechnic Institute

Stardust

It is worth showing that the composition of a human is very different from the universe as a whole. Of every 10,000 atoms in the Sun, 7400 are H, 2440 are He, 3 are C, 2 are N, 5 are O, and 150 all other elements. Of the same 10,000 atoms in the human body, 6500 are H, 2 are He, 2000 are C, 500 are N, 900 are O, and 100 all other elements. Our bodies are therefore cinders formed in the residue of a stellar collapse in a universe that is overwhelmingly H and He. Stress that the cinders came from a previous generation of stars. And that all the elements we crave on our planet (silver, gold, platinum) were fried in the death throes of stars.

Chris Impey
University of Arizona

Stellar Lives

Use car or money analogies to get across ideas of slow or fast stellar evolution. Dwarfs or low mass main-sequence stars are poor misers, with not much money but eking it out over a long time. Or Japanese cars, with small fuel tanks but gas-efficient engines. Giants and high mass main-sequence stars are then spendthrifts winning a fortune on the lottery and spending it in a week, or drag racers with a massive fuel tank and a thirsty engine that can suck it dry very quickly.

Comment on the philosophical implications of our theory of star formation: that from the laws of physics, a theorist could deduce the existence, mass range, nuclear burning and shape of stars (even if we lived on a shrouded planet where no telescopes could function).

Chris Impey
University of Arizona

Stellar Evolution by Inquiry

One of the most difficult concepts to teach in astronomy without resorting to rote student memorization is the life cycles of stars. Developed through considerable trial and error, we would like to share our student-centered, learning cycle approach to teaching star life cycles. The theme of the stellar evolution inquiry lesson is based on the observational nature of astronomy. Because the lifetimes of stars are extremely long, ranging roughly from 40,000 years to more than 12 billion years, astronomers have never actually observed a single star go from "birth to death." Yet, astronomers have a reasonable feel for how stars change over time. Analogously, our students have never observed a single human being go through a natural 70 year progression from birth to death. Despite students' brief time on Earth, our students do understand how human beings change from being an infant, to a toddler, a child, a teenager, a young adult, middle age, senior citizen, and eventually, death, even though they have never observed this on a single adult. It is this strategy of inferring patterns of change that we use to teach students about the star life cycles.

We begin the unit with a two-part exploration. We first provide students with a random stack of pictures of different friends and relatives of various ages. We challenge them to arrange the random picture in sequence from the youngest person to the oldest person. They can usually accomplish this task with minimal discussion although we could increase the difficulty easily. We then ask students to brainstorm how they could obtain pictures showing every possible step in the human life cycle. Common suggestions are to look at large family reunion photographs, review school yearbooks, or visit hospitals; anywhere that one would find many people at different stages in their life cycle.

The second part of the exploration is to provide students with a set of Hubble Space Telescope pictures and we ask them to sequence them from youngest to oldest. Given the limited background in astronomy of most of our students, this is indeed a challenge, even when we let them utilize the full-resources of the Internet. We require students to provide written rationale for why they believe one image shows a more mature star than other images show.

On the second day of the lesson, we begin to provide students with an introduction to the concepts of star life cycles using an interactive lecture strategy. We use flow charts as concept maps to describe the progression of stars from stable main sequence stars to red giants and to the various end-states depending on the stars' initial masses. These exotic end-states can include white dwarfs, neutron stars, and black holes. We use the same images that students used in their exploration and add labels showing accurate vocabulary. Such a strategy allows me to make explicit connections between the images and the words that students use with scientific vocabulary. Eventually, we'll use student constructed concept maps as a portion of their unit assessment.

On the last days of the lesson, we challenge students to apply their knowledge about star life cycles. Teams of students are required to create product "sales brochures" for either low or high mass stars and make a convincing "sales presentation" to class members that accurately describe the life cycle of their assigned star type. We believe that students find this part of the lesson the most challenging as well as the most entertaining. As part of an embedded assessment, students are required to critique the presentation and materials for scientific accuracy. This three-stage learning cycle approach combines effective instruction with engaging NASA pictures and student-centered work. Certainly, not every aspect and theoretical detail of stellar evolution is included in the lesson; however, it is the enterprise of scientific inference through experiment and observation that we hope our students remember in the distant future.

Jeff Adams and Tim Slater
Montana State University

Chapter 14
Galaxies and Cosmology

The Scale of the Local Group

Think of the Milky Way Galaxy as a dinner plate parallel to the floor with a baseball stuck through the center. Then M31, the Andromeda Galaxy, is a slightly larger plate with a similar baseball stuck in it over on the other side of the room, somewhat farther off the floor, inclined so that the Milky Way plate is about in its plane. The Magellanic Clouds are cotton balls about two plate diameters below the Milky Way, and M31's companions are two cotton balls, one in front of the other behind M31. M33 is a dessert plate, with a marble stuck in it, several diameters to the right of M31, inclined so the Milky Way is directly above its plane. Cotton balls of different sizes scattered throughout the room represent the other, smaller members of the Local Group.

Anthony J. Weitenbeck
University of Wisconsin

Hubble's Law and the Expanding Universe

The typical way to discuss and demonstrate Hubble's Law and the uniform expansion of the universe is by drawing on the chalkboard using the raisin cake model. A better and more visual way is to construct a three-dimensional model of the universe from small Styrofoam or clay spheres for galaxies and dowel rods connecting them in a 3-D array. Even marshmallows and spaghetti sticks can be used for the model. Place about 9 spheres on a flat surface and connect one to the other in a 3x3 array using the rods. Them make two more arrays just like this one and finally mount these two on the original in a stacked manner forming a 3x3 array in each dimension with equal spacing between each galaxy. This is not critical, but the model will look better if equal distances are used. Now construct another 3-D array that has dimensions twice or three times the first one. This represents the universe at a later time when it has expanded.

Show the students the first array and explain the representation. Which galaxy are we? It won't matter since the center of the universe cannot be located anyway. Choose one and measure the distances to several other galaxies with a meter stick or string. Distances can be measured in all directions. Now show the larger array and comment on the expansion of the universe and its uniformity making the appearance larger but similar in arrangement of galaxies. Now measure the distances from the same reference galaxy to the others on the same relative locations in the expanded version. A record of distances before and after can be used to find the distance moved by each.

A plot of original distance on the x-axis and change in distances in equal time on the y-axis will give a straight line, the slope of which is Hubble's Constant for this universe. Several features of the model can be related to the actual expansion of our real universe. The farther galaxies receded faster, each equal step out in space results in an equal increase in recessional velocity, and the selection of any galaxy as the reference one will result in the same law with the same slope. Try it.

James E. Kettler
Ohio University - Eastern Campus

The Cosmic Calendar Revisited

Carl Sagan's "Cosmic Calendar," presented in his COSMOS series, was a hit, but few of us have Carl's special effects budget! And I thought that the commercial poster version somewhat lost the calendar analogy. My more modest Cosmic Calendar is literally a calendar; one of those large thirty- by forty-inch flip calendars (one sheet per month) sold by office supply companies. (It's big enough for a whole class to see.) As in COSMOS, I scale the entire history of the Universe, from big bang to the present, into one calendar year. Thus, "big bang" is printed with large letters in the box for January 1. Now is December 31. I have included other significant "dates" in between, such as the formation of the Milky Way and the birth of the Solar System. I discuss each as I rapidly leaf through the pages. Many early months are blank, of coarse. These pages contrast with relatively "busy" later months that include events such as the first appearance of mammals, primates, and human beings. Students take note of the fact that we do not show up until New Year's Eve, and that all recorded history takes place in the last minute of the last day of the cosmic year. What, I ask, will the next minute bring?

Thomas Hockey
University of Northern Iowa

Galactic "Baseball Cards"

When I introduce the subject of galaxies external to our own, I ask students to notice that we are examining the same short list of observable properties with galaxies and clusters of galaxies that we did with planets and stars; distance, size, mass, luminosity, motion, and appearance. As with most scientific classification schemes, we start with appearance. I photocopy sets of pictures of galaxies for each class member, and as an exploratory exercise, ask the students to come up with classification systems of their own. (I try to make the images the same size and brightness, so that no other information but shape and appearance is provided by the pictures.)

Many students produce on their own a system that parallels that of Edwin Hubble--spirals, ellipticals, and irregulars. (They call them things like "pinwheels," "blobs," and "other," but the idea is the same.) When then introduced to Hubble's work, it seems natural and obvious to many. Other students put some spiral galaxies seen edge-on into the same category as they do some highly oblate ellipticals; this provides an opportunity to discuss how the ambiguity of viewing geometry makes interpreting galaxies more complicated. Still others put some face-on spirals, with tightly wound arms, together with nearly spherical ellipticals. Clearly, they see, while a good starting point, visual appearance does not take us all the way to an uncontroversial classification scheme. Other properties (such as stellar population and the presence of gas and dust) must be brought into the game.

Thomas Hockey
University of Northern Iowa

Mapping the Milky Way; A Metaphor

Mapping the Milky Way Galaxy has been a triumph of twentieth-century astronomy, prevented as we are, by time and space, from ever venturing beyond our galaxy's limits. I compare the job to drawing a picture of the outside of a student's house from the inside. Outside, the house appears as brick, trim, and roof. Inside, it is walls, ceiling, and furniture. The two pictures are very different. Yet with insight drawn from the appearance of other houses seen out the windows, and careful attention to dimensions and geometry, the student can imagine coming up with a plan of what the exterior might look like, even if he or she could never step out the front door to take a look.

Thomas Hockey
University of Northern Iowa

The Anthropic Principle

Although speculative, this subject introduces powerful physical and philosophical concepts. Discuss premise that fundamental parameters of our universe may be connected with our presence as observers. Note that our microscopic theory of quantum mechanics gives enormous importance to the role of the observer. At the cosmological level, both the enormous size and age of the universe may be connected with our existence as observers, since a universe with very different parameters would be "stillborn" as far as intelligent life goes. Basically, any low-density universe would expand so quickly that gravity would not form stars and galaxies, and so nucleosynthesis would not form the heavy elements on which life depends. Any very high-density universe would recollapse into a "Big Crunch" before enrichment cycles in stars had deposited heavy elements in the interstellar medium, and hence into planets and life. Unfortunately, the anthropic principle may just be a logical curiosity, since it is post hoc and makes no predictions.

Chris Impey
University of Arizona

Broken Symmetry in Cosmology

Good analogies are hard to find for the esoteric phenomena of the early universe. A key element of modern physics is the concept of symmetry. Experiments at the microscopic level reveal a complete equivalence between particles and their partners with opposite quantum properties, the antiparticles. In the very early universe matter and antimatter were in equilibrium with the intense radiation field, and there were equal amounts of each. Yet we currently find ourselves in a universe that is overwhelmingly composed of matter. How did this happen? An underlying symmetry that is clearly seen at high temperatures may be broken at lower temperatures. The obvious example is water and ice. Water molecules are found with all orientations in the liquid state, but at the freezing point they line up to form the crystals and fissures that represent the broken symmetry of ice. Magnetic domains aligning at the Curie temperature is another example. Perhaps the easiest analogy for students to grasp is the spinning coin. A spinning coin obviously has more energy than a coin at rest, yet it displays no preference for heads over tails. Only when the energy is lowered must the coin break its symmetry and choose one of the two possible states. This makes an obvious analogy with the broken symmetry that led to the dominance of matter over antimatter in our universe.

Chris Impey
University of Arizona

Microwave Background Radiation

To the untutored eye, the universe appears to be mostly composed of matter. The structure is dominated by galaxies and the stars they contain, plus a large amount of dark matter. Yet there are over a billion photons for every particle in the universe. The radiation that was so important in the early phases of the big bang has been redshifted by the expansion of space to a very low energy. If you draw a wave on a balloon and inflate it, students can see that the expansion of space leads to an increasing wavelength and a decreasing energy for radiation. Nevertheless, the faded glory of the early universe is still around us. There are a billion photons of the microwave background in every cubic meter of space. Point out that there is no health hazard, the combined power is only 10^{-5} watts.

Chris Impey
University of Arizona

The Concept of Horizons

Students are often perplexed by the concepts of cosmology. One distinction that is particularly important and usually misunderstood is the difference between the observable universe and the physical universe. This is a consequence of the finite speed of light and the finite age of the universe. We can only see regions of the universe from which light has had time to reach us, and we see those regions as they were in the distant past--the concept of lookback time. However, depending on the best value for the age of the universe, we can only see regions that are about 15 billion light-years away. Light has not had time to reach us from more distant regions. This sphere is the observable universe, its boundary is the horizon. Every day, light from regions 1 light-day further away enters our horizon and reaches us for the first time. It is not a physical limit, just an information limit. The entire physical universe may be much larger, in fact in the inflationary cosmology the observable universe is a tiny fraction of the physical universe. Beyond the horizon, regions are moving away from us faster than the speed of light in the expanding universe. Special relativity only governs the passage of signals within space, and so its limitations do not govern this situation where the superluminal motion is caused by the expansion of space itself.

Chris Impey
University of Arizona

Inflationary Cosmology

To appreciate the benefits of the inflationary cosmology, have students imagine a small balloon that is inflated by a prodigious amount, say from the size of a grapefruit to the size of the Earth. Any curvature that the balloon originally had would be drastically reduced (just as the Earth appears flat to a person restricted to a small portion of it). This is the answer to the so-called "flatness" problem: the fact that the universe has a close-to-flat geometry now, which would have required extraordinarily fine-tuned conditions in the very early universe.

Chris Impey
University of Arizona

Gravitational and Inertial Mass

Einstein was led to the general theory of relativity by pondering what seemed to be a coincidence in physics. The inertial mass of an object is the resistance it presents to a change in its motion. The gravitational mass of an object is its response to a gravitational field; in other words, its weight. The two are measured to be identical, yet they appear to be very different forms of mass. Imagine a large, smooth piece of iron resting on a slick, ice surface.

The gravitational mass dictates the force with which the iron presses down on the ice. The inertial mass is the resistance of the iron to a change in speed, either trying to speed it up or slow it down. This has nothing to do with gravity, since the motion on the ice is horizontal and gravity acts vertically. Moreover, if the iron was in orbit around the Earth, it would weigh nothing and so have zero gravitational mass, yet its inertial mass would not change. Inertial and gravitational masses are equal to an exquisite degree of precision: modern experiments find the difference to be less than 1 part in 10^{15}. To Einstein, this was more than a coincidence; he believed that it held the key to understanding.

Chris Impey
University of Arizona

Escape Velocity Analogy for Expansion

The best analogy for the expansion of the universe uses the idea of escape velocity. Consider a rocket launched from the surface of the Earth. A rocket launched with an initial velocity of less than 11 km/s will slow down or decelerate as it rises. The Earth's gravitational attraction will eventually overcome the upward velocity, and force the rocket back to the planet's surface.

On the other hand, a rocket with an initial velocity above 11 km/s will escape from the Earth forever. Of course, the rocket will continue to slow down since the planet's gravity has a long reach, but it will never reverse its direction and fall back to Earth. In our universe, the Hubble parameter is analogous to the launch velocity of the rocket. The mean density is analogous to the mass of the planet. For each possible launch velocity of a rocket, there is a mass of planet where that velocity will equal the escape velocity. Conversely, for every planet there is a single velocity that will allow the rocket to just escape gravity's pull. Larger planets have larger escape velocities. By analogy, for every possible value of the Hubble parameter, there is a mean density that will just be sufficient to stop the universal expansion and close the universe. A larger Hubble parameter requires a larger mean density to close the universe. Both quantities can, in principle, be measured.

Chris Impey
University of Arizona

Selection Effects in Deep Galaxy Surveys

Deep surveys tend to preferentially select galaxies that are particularly luminous and blue. Biases of this sort are called selection effects. Cosmological tests often require samples that are complete within a particular *volume*, because all galaxies in a volume must be counted to calculate the mean density. On the other hand, a deep search for galaxies will be limited to a particular *magnitude*.

The distinction can be made clear with the following example: suppose that any particular volume of space has four times as many dwarf elliptical galaxies as large spiral galaxies. Suppose also that a large spiral is four times as bright as a dwarf elliptical. Because spirals are more luminous than dwarf ellipticals, they can be seen out to a larger distance. If we search the full available volume, we find 20% of the *number* of galaxies to be spirals, but since spirals are four times as bright as dwarf ellipticals, 50% of the *total light* is contributed by spirals. However, if we instead search down to an apparent magnitude limit the result is quite different. Spirals can be found out to twice the distance, or eight times the volume. Therefore, we find 70% of the *number* of galaxies to be spirals (8 spirals and 4 dwarf ellipticals), and 90% of the *total light* to be contributed by spirals (because they are twice as numerous and four times as luminous). Rare and luminous galaxies are overrepresented in any magnitude limited search. Exactly the same effect can be seen in the solar neighborhood. The *brightest* stars in the sky are mostly giants and hot main sequence stars. However, a full census of the volume around the Sun shows that the *most common* stars are dwarfs.

Chris Impey
University of Arizona

Rubber Band Hubble's Law

Paste some glitter stars on a wide rubber band. As you stretch the rubber band, note that the more distant the stars are from one another, the faster they move. Extend the analogy by pointing out that if the glitter stars were aphids, their crawling along the rubber band would induce intrinsic local velocity effects similar to those which make Hubble's constant difficult to determine from observations of nearby (say < 10 Mpc) galaxies, whose distances are more certain. Point out that drawing stars on the rubber band isn't a good analogy, because as the rubber band stretches the stars enlarge, but as the universe expands, galaxies stay the same size. (This point was also brought home in Woody Allen's movie Annie Hall when the schoolboy Woody was told by his psychiatrist not to worry about the expanding universe because, although that may be true, Brooklyn was not expanding and the young Woody should get back to his studies.) The same analogy may be done with a balloon.

John Broderick
Virginia Polytechnic Institute

(Galactic) Clouds in My Coffee

I have always marveled at how easy it is create a spiral galaxy in a cup of coffee, even if I'm half asleep. All I need to do is stir the coffee, and add a little cream. I can create some very good replicas of the different Hubble types.

Why is the fluid flow in a coffee cup nearly identical to the fluid flow on a much larger galactic scale? The answer turns out to be the same reason we can build wind tunnels and test tiny model airplanes in order to design the full-scale one. Fluid flow models are valid if they have the same Reynolds number, which is a number which incorporates many properties of fluids. The Reynolds number of a fluid is determined by things like its density, temperature, velocity, and viscosity. The coffee and cream combination and spiral galaxies have the same Reynolds number, even though the parameters that go into the Reynolds number are all different for the two situations.

Thus, the coffee is a good simulation of the large scale fluid flow. For a more detailed discussion see P. Stevens wonderful book Patterns in Nature. (I have slides that illustrate this phenomena very well.) Students really enjoy these visual metaphors that help unify cosmic structures with ones on a human level. Concepts such as galactic shear, spiral arm windup, or the von Weizsacker theory for the proto-solar nebula can all be illustrated by the cream in a coffee cup. Plus, it gives you a good excuse to drink coffee in class.

Stephen M. Pompea
Pompea & Associates

A Galaxy: A Dissection

I throw out a teaching challenge. In this era of multi-wavelength measurements, with hundreds of great slides on every aspect of extragalactic astronomy , I challenge you to simplify. It is often tempting to cover many aspects of galaxies on separate days, and the effect is often a disjointed presentation. I challenge you to base an entire class session on only one visual aid, a color slide or poster of one nearby galaxy. The picture I am thinking of is on the cover of Tim Ferris's great visual extravaganza Galaxies (1982, Stewart, Tabori and Chang, New York) and is of the galaxy NGC 6744 (M83). Actually, any detailed color photograph of a nearby galaxy would do. The photograph is remarkable enough that you could look at it for hours and always see more. Students need to look at pictures like this and try to understand all of the different processes going on in a galaxy. Many processes are delineated by their color: the reds of HII regions, the yellow of the bulge population, the blues of young star complexes.

This process of analyzing a single complex object in front of them is similar to biological dissection. In biological dissection it is the student's job to develop the keen eye necessary to differentiate the component parts. It is also up to the student to develop the concentration necessary to follow a particular line of thought or observation. For example, a simple question like: "How many spiral arms are there?" does not always have a simple answer. The process of arriving at a thoughtful understanding is important.

Stephen M. Pompea
Pompea & Associates

Quasars

Students have difficulty grasping the enormous power and compactness of a quasar energy source, and the fact that although they appear stellar, they are situated at the nuclei of galaxies. Consider the view of Los Angeles at night as seen from a helicopter hovering high above the city. The area has over 100 million lights; consider each one a star in a normal galaxy. Then a quasar is a light source with the power of 1000 times all the lights in the city, concentrated in a 1-inch area at the center of the city. As you fly away from the city, the intense central light is visible long after the individual house and street lights have faded from view. Note that even if we don't understand the power source in quasars, we can use their intense light to study the vast regions of the universe between us and them. Their use as searchlights to illuminate cold and dark material that could never be seen directly is important.

Chris Impey
University of Arizona

Early Universe

Stress predictions of Big Bang model, microwave background and nucleosynthesis. Story of Penzias/Wilson discovery is nice, that they nearly published an erroneous early detection of the microwave background due to a "thin, white dielectric film" coating the surface of their radio telescope (it was pigeon dung). Use dominance of matter over antimatter in our universe as an introduction to the idea of symmetry breaking. Use example of a spinning coin, which when it falls must choose between two options that are equally probable when it is spinning. Note that in the early universe, it is the enormous temperatures that create the symmetry between the forces of nature.

Chris Impey
University of Arizona

Chapter 15 Astrobiology and the Search for Extraterrestrial Life

The Message

Students can send a message to each other to give them a feel for the difficulty of constructing or decoding messages. The simplest message might have the following hints, which you may choose not to share with the students. The message could be pictorial, and be constructed in pixels, or picture elements. Each pixel is like a square on a piece of graph paper. The square may be blank or it may be colored in. There may be a symbol in the message that is the equivalent of the carriage return on a typewriter, i.e., to start a new line.

Each student gets to construct a message, some form of picture that can be drawn using the squares on a piece of graph paper with lots of little squares. The message should try to communicate something of importance to an extraterrestrial civilization. For example, ask your students to try to show "them" that we know mathematics (trigonometry, algebra, calculus?). The message should contain information about our solar system. Perhaps we want to keep our exact location secret, to protect ourselves from violent outsiders. Good discussion point. If so, why are we sending a message? At the next class session pick one message to communicate to the class. A filled-in square could be represented by the word "beep," a blank square by a clap of the hands, and a new line by a pound on the table, so the message would be an acoustic string.

For a more sophisticated class or for a take-home exercise, give the class a string of 1s and 0s to represent the same thing. Let them figure out the shape of the picture. Make the number of data bits be divisible by two prime numbers, the dimensions of the array. Let the students try to interpret the picture, once they have the whole message. Even with no transcription error, it may be difficult to recreate the thought process of the sender. Students may wonder why the message must be sent in such a primitive fashion, since we seem to get a whole television picture all at once. Now is a good time to tell them about data bits, and how pictures are really transmitted (see related idea on pictures and data bits).

This is a fun puzzle for the class. The instructor can design some rather esoteric messages and give them to the class in raw or final form. This exercise really gets the class thinking about our culture and how we think.

Stephen M. Pompea
Pompea & Associates

Time Capsules

This exercise raises some important questions for our students on what is unique and valuable about us as humans. The Pioneer 10 and 11 spacecraft had a plaque showing the solar system and which planet the spacecraft came from. The Voyager 1 and 2 spacecraft had a video disk with assorted information about life on earth. When teaching about the possibility of extraterrestrial life, it is worthwhile to ask the students to think about the equivalent of a spacecraft time capsule. What information about us as humans is important enough to communicate across the vastness of space. Suppose human life on earth ends and the spacecraft time capsule is all that will be known about the inhabitants of planet number 3? What would your students want to communicate about life on earth? What media best can express information about ourselves?

Stephen M. Pompea
Pompea & Associates

Teaching about UFOs

I have fun teaching about UFOs, in part because there is such intense interest from the public, and so much fascinating physics/astronomy that can be taught at the same time. I start by stating that much of the problem is the lack of detailed investigation which comes with UFO reports, combined with the common ignorance most people face of the heavens, and in addition, we discuss witness reliability. Then I spend time discussing atmospheric optics, which can give the students an idea of the beautiful diffraction and reflection and refraction effects which are common in the atmosphere but hardly ever noticed. After these lectures, the students are amazed with how much better they understand the sky--both in the day and at night, and nearly daily someone will come in with a new "sighting" of something they *can* explain.

> Discuss problems with angular velocity as opposed to really knowing distance or height. It is good to dig out various explained and unexplained UFO reports, and to point out how often several witnesses describe similar events completely differently. Especially good examples are flying aircraft, which typically have the wrong number of objects described, speed and direction.

> Use UFOs to illustrate parallax and how people claim that Venus can chase them. Note that many air traffic controllers have "cleared" Venus for landing!

> Psychological effect (optical) of the vault of the heavens being flattened--leads to gross misestimation of angles. A good reference discussing this is found in Minnaert (1954).

<u>Witness Reliability</u>

A good example which indicates both how modern people tend to make poor witnesses, and how incredibly unaware people are, is to get into a discussion about archaeoastronomy, by pointing out how well ancient peoples understood the motions of the celestial phenomena which occurred both on long and short timescales. Astronomical alignments in ancient ruins throughout the world serve as ancient testimony to the sophistication of knowledge achieved by simple observation--a talent which has generally been lost. Simple labs can be constructed for the students which illustrate how the early civilizations might have made these measurements, and in addition how to ask the right questions to debunk the pseudoscientific acclaims of alignments which do not exist.

In particular, in Hawaii, I have been working with a retired architect who has made some amazing discoveries about the sophistication of the ancient Polynesian navigators and their astronomy, by tying together the ancient hula chants with astronomical alignments along solstice lines between islands in the Hawaiian chain. I feel it is always important to create interest in the subject being taught by flavoring the science with local knowledge and heritage.

One of the problems with UFOs is that in order to fully understand what is going on, an investigation as thorough as any murder investigation must be undertaken, and there are rarely resources for such a task. Like a murder investigation, the psychology of the witness is also an important factor which must be considered, and there are simple exercises which can show the students how poor people usually are as witnesses. A fun exercise--much like the telephone game as a kid--have each student look at a somewhat complex figure for 10 sec and draw it. Each successive student draws from the previous student's art.

Karen Meech
University of Hawaii

UFOs and Atmospheric Optics

There are a myriad of atmospheric phenomena which occur frequently, which most city dwellers are completely unaware of because they rarely take the time to look at the sky around them. On the rare occasion in which they are spotted, they can be the source of UFO reports. While rainbows are never misinterpreted as UFOs, they are a good place to start in terms of the simplicity of explaining atmospheric phenomena. In particular, the phenomena which are less common and have been mistaken in the past for UFO phenomena include various ice refraction reflection effects, such as: ice haloes (corona), fog and moonbows, parhelic arcs, sun pillars, and spectre de brocken.

Fascinating demos can be done simply on rainbow formation, corona, the greenflash, etc., in the classroom. They look very impressive, in part because I use dry ice for fog (corona). Good experiments are found in Bohren (1987, 1991). A good introduction to physics is found in Greenler (1980). Along the lines of discussing unusual phenomena, the occurrence of ball lightening its relation to geophysical (tectonic) phenomena is also very interesting to discuss. This is not something that most students will be familiar with. References on this subject are few and far between, and most that I have come across have been in *Nature*.

Bohren, C. F. (1987). "Clouds in a Glass of Beer. Simple Experiments in Atmospheric Physics," John Wiley & Sons, Inc., New York.
Bohren, C. F. (1991). "What Light Through Yonder Window Breaks? More Experiments in Atmospheric Physics," John Wiley & Sons, Inc., New York.
Greenler, R. (1980). "Rainbows, Haloes and Glories," Cambridge University Press, Cambridge.

Meinel, A. and M. Meinel (1983). "Sunsets, Twilights and Evening Skies," Cambridge Univ. Press, Cambridge.
Minnaert, M. (1954). "The Nature of Light and Color in the Open Air," Dover Pub., Inc. New York.

Karen Meech
University of Hawaii

Listening for E.T.

Students need only look at a portable radio receiver (which I bring to class) to see the challenge of the SETI program. To receive a radio program clearly, the antenna must be adjusted properly. The correct station also must be selected. When I turn the radio on, it has been pre-turned so that nothing will be heard. I then turn through the frequencies slowly--my radio still has a nice big dial for this purpose--until I begin to pick up a station. A twist of the antenna brings it in static-free. On my tour of the FM spectrum, I point out that at any time we might be "pointing" at an intelligent signal (the antenna is positioned to receive it), but we are on the wrong channel or visa versa.

In the real search for extraterrestrial radio signals, of course, the number of possible directions and frequencies is fantastically greater than in my analogy. (Only a computer could point and tune fast enough.) And we have no map to tell us what is the correct direction and no *TV Guide* to tell us which is the correct frequency. (Parenthetically, a scan of the commercials bands from my campus often results in a class discussion of what kind of "signal" indicates truly intelligent life!)

Thomas Hockey
University of Northern Iowa

Uncertainty and the Drake Equation

To show the uncertainty of the Drake equation, calculate high and low estimates of all the factors, emphasizing increasing uncertainty as you go from astronomical and geological factors to biological and then sociological factors. Resulting range of stars in our galaxy with intelligent life is < 1 to several billion; corresponding distances are 10^7 to 15 light years. Emphasize the dominance of anthropocentric arguments in this subject (the Pioneer 10 plaque can be used to make this point humorously). Mention that although we consider life elsewhere in the solar system unlikely, we have recently discovered complete food chains deep in the oceans that metabolize H_2S and live in an environment that is hotter and under more pressure than the surface of Venus, and that we have found life deep in the Arctic ice pack, an environment more barren than the surface of Mars.

Chris Impey
University of Arizona

What Is Important to Us?

This exercise raises some important questions for our students on what is unique and valuable about us as humans. The Pioneer 10 and 11 spacecraft had a plaque showing the solar system and which planet the spacecraft came from. The Voyager 1 and 2 spacecraft had a video disk with assorted information about life on earth. When teaching about the possibility of extraterrestrial life, it is worthwhile to ask the students to think about the equivalent of a spacecraft time capsule. What information about us as humans is important enough to communicate across the vastness of space? Suppose human life on earth ends and the spacecraft time capsule is all that will be known about the inhabitants of planet number 3. What would your students want to communicate about life on earth? What media best can express information about ourselves?

Stephen M. Pompea
Pompea & Associates

Designer Aliens

Here is an exercise that helps students integrate the knowledge they have learned in their class about stars, radiation, blackbodies, surface gravity, and planetary atmospheres. The students must design a life form that can prosper on a planet. The students are given a type of star (say F8), a planet of a certain radius and density, and of a certain distance from the star. The student is also told how much time has elapsed after the birth of the star. The student must determine if the star is still around, if the planet is habitable, and what the surface gravity conditions on the planet would be. Also, assuming that a potpourri of gases are released by volcanoes, would any gases useful to life as we know it remain in the atmosphere? To do this, the student must have an understanding of the spectrum and intensity of the radiation from the star, and tie this in with the concepts of escape velocity.

If life can exist, what forms might be most successful? It is not necessary that detailed calculations be made; many of the factors can be reasoned through by analogy with Earth, if the concepts are well understood. For example, if the planet is half as far from a star that puts out 1/3 times as much light as our sun, and the planet is twice as dense as the earth, and is half the radius, what will the temperature be like, compared to the Earth? The approximate answer can be reasoned through. If the temperature is known, which atmospheric constituents can be retained?

This is a fun exercise for students of all levels of knowledge and sophistication. It can even be used to illustrate the basic concepts themselves, such as the inverse square law. It is also an interesting background exercise if you want to discuss and evaluate the Gaia hypothesis. Students could also draw the planet's surface and the life forms adapted to it.

Stephen M. Pompea
Pompea & Associates

Planetary Zone of Habitability - The Goldilocks Problem

Do the new planets discovered around distant stars harbor life? Every planetary system surrounding a main-sequence star will have a region in which water could be found in the liquid state. This region is often called the Zone of Habitability and planets or moons orbiting in this zone have the potential to support life. We guide students to conceptualize the important parameters that define the size and location of this zone for a particular star, including luminosity, spectral class, planetary orbital radius, and existence of planetary atmospheres. As a first-degree approximation, students examine Venus, Earth, and Mars where they find that Venus is "too hot", Mars is "too cold," and Earth is "just right."

This classification is reminiscent of the story of Goldilocks and the Three Bears and serves as a departure point for discussing the search for life on extra-solar planets.

Donna Governor
Brown Barge Middle School, Pensacola, Florida

Allison Simmerman
Honoka'a, Hawaii

Ed Prather
Montana State University

Extrasolar Planets

Finding extrasolar planets and doing research on them is a rapidly expanding field, and is a good area for students to watch evolve. Textbooks will have a hard time keeping up with the latest discoveries and research in this area. Indeed, two monthly science magazines proclaimed in the same month's (March 2000) issue on their covers that 29 and 33 extrasolar planets had been discovered. The web is a great resource for keeping up with the latest discoveries. Two of the best sites are:

Extrasolar Planet Search Project at San Francisco State University
http://www.exoplanets.com/

Extrasolar Planets Encyclopedia. This website at the Observatoire de Paris is a comprehensive encyclopedia of all aspects of extrasolar planet research.
http://www.obspm.fr/encycl/encycl.html

Stephen M. Pompea
Pompea & Associates

Searching for Extrasolar Planets Using the Doppler Effect

Although it is currently impossible to directly image extrasolar planets, it is possible to indirectly detect their presence by examining a star's Doppler shift. Unseen orbiting planets cause the star to oscillate back and forth or "wobble" which results in the starlight being Doppler shifted. We provide students with actual velocity versus time graphs from recent publications and challenge them to determine both the orbital period and orbital radius for unseen planetary companion and ask them to make comparisons to planets in our own solar system.

Ed Prather
Montana State University

Sandy Shutey
Butte High School, Butte, Montana

Marty Wells
LaSalle High School, Cincinnati, Ohio

Environmental Impacts of Interplanetary Exploration

Humans have long looked towards distant planets with the hope of establishing frontier colonies. However, our growing knowledge of destructive human impacts on our own planet makes us cautious about how we undertake our interplanetary explorations. In an effort to help students appreciate the interrelationships between extraterrestrial ecology and human endeavors, we challenge students to develop a protocol for minimizing the impact of our exploration. We first show students Viking and Pathfinder photographs of the Martian surface and ask them how these robotic devices might be impacting Martian ecology. Students watch video clips and then summarize Ray Bradbury's *Martian Chronicles* to initiate discussions about the effects of human exploration. Finally, we ask students to write brief public policy statements to guide future space missions.

Mel Anderson
Pittsburg Middle School, Pittsburg, Kansas

Janet Erickson
C. R. Anderson Middle School, Helena, Montana

Ed Prather
Montana State University

Martian Soil Simulant

A Martian soil simulant has been developed for use in scientific research, engineering studies, and education. It is called JSC Mars-1, and is a natural material obtained from volcanic ash. It has various properties which are similar to Martian soil. These properties include its reflectance spectra, mineralogy, chemical composition, density, grain size, porosity, and magnetic properties. The material is formed from volcanic ash from Mauna Kea, Hawaii.

According to the August 25, 1998 issue of EOS, Transactions of the American Geophysical Union, small samples of the soil simulant were to be made available for distribution to investigators and educators. The samples are free, with recipients paying the shipping costs. At that time, samples could be obtained by contacting the Office of the Curator, NASA Johnson Space Center, Houston, TX, 77058 USA.

Stephen M. Pompea
Pompea & Associates

Teaching Evolution in the Science Classroom

Evolution is a fundamental concept in science: systems change over time. Change is evident in physical systems: planetary systems form, stars evolve, continents drift, sediments deposit and erode. In biological systems, evolution is the fundamental theory through which scientists explain life. Premier biologist, Theodosius Dobzhansky (1900-1975) wrote, "Nothing in biology makes sense except in the light of evolution." (American Biology Teacher, Volume 35 pages 125-129, 1973). Yet, teaching evolution in American public school classroom stirs controversy. Teachers are pressured to ignore evolution or disguise it with euphemisms like "change over time." In some states, disclaimers are printed in textbooks.

Public attitudes toward the teaching of evolution surprise many scientists and science teachers. An analysis of Gallup Poll results reveals the public's attitude:

Which of the following statements comes closest to your views of the origin and development of human beings?

	Aug. 1999	Nov. 1997	June 1993	1982
Human beings developed from less advanced forms of life, but God guided the process.	40%	39%	35%	38%
Human beings have developed over millions of years from less advanced forms of life. God had no part in this process	9%	10%	11%	9%
God created human beings pretty much in their present form at one time within the last 10,000 years or so.	47%	44%	47%	44%
No opinion	4%	7%	7%	9%

The scientific explanation for the origin and evolution of life and humans is not widely accepted by the public even though the theory of evolution is fundamental to our understanding of the natural world. This attitude impacts all of science. In astronomy courses, the origin of the universe and the evolution of stars and planets are commonly taught. The evidence is gathered and presented with the presumption that students accept this evidence and our conclusions as a true description of the natural world. They may or may not, depending partially upon initial attitudes as well as the quality of the class.

Commonly, evolution of the universe and life are dismissed as "just theories" by opponents to evolution, not understanding that in science a theory rests firmly on a foundation of fact. This is a misunderstanding of the nature of science. In astronomy, the theory of the origin of the universe (the Big Bang) is recommended in both the National Science Education Standards and The Benchmarks for Science Literacy, yet it appears in just over half (54%) of the state frameworks. The same is true for concepts related to the formation of the solar system.

Resources and materials are available to assist educators in teaching about evolution and the nature of science. Useful ones are:

American Association for the Advancement of Science, Benchmarks for Science Literacy, Oxford University Press, 1993. (also on CD-ROM)

Lerner, Lawrence. State Science Standards: An Appraisal of Science Standards in 36 States, Fordham Foundation, Washington DC, 1998. (also online)

National Research Council, National Science Education Standards, National Academy Press, Washington, DC 1995 (also online)

National Academy of Sciences, Teaching About Evolution and the Nature of Science, National Academy Press, Washington, DC 1998 (also online)

National Academy of Sciences, Science and Creationism, Second Edition, National Academy Press, Washington, DC 1998 (also online)

Skehan, James W. Modern Science and the Book of Genesis, NSTA, Arlington, VA. 1986 (also online)

On the Web

Database of State Science Standards
http://www.achieve.org/achieve/achievestart.nsf?opendatabase
-assembled by Achieve, a non-profit organization concerned with the quality of American schools; can be searched.

National Science Teachers Association home page
http://www.nsta.org
-leading science teacher membership organization; publications & conferences; up-to-date information on evolution in education

National Academy of Sciences evolution web site:
http://www4.natioanlacademies.org/opus/evolve.nsf
-national scientific leader, publications & conferences

National Center for Science Education:
http://www.natcenscied.org/
-non-profit center for evolution education and advocacy

NASA Space Sciences Program:
http://spacescience.nasa.gov
-your space agency ☺

SETI Institute
http://www.seti.org
-non-profit scientific research and educational projects based upon the search for life beyond Earth

Edna DeVore
SETI Institute

Finding Extraterrestrial Life from Space

Finding "life" outside our own solar system requires a detailed understanding of the signatures of extant life on Earth. The hot, cold, dark, acidic, and dry extreme environments in which life surprisingly exists on Earth also exist on distant planets and moons. Student examine actual images from planetary space probes of Earth, Mars, Io, Europa, and Titan with the goal of finding likely environments in which life may exist. Students are challenged to propose future missions throughout the solar system that would have some chance of finding life.

Buck Buchannon
Belgrade Middle School, Manhattan, Montana

Ed Prather
Montana State University

Course on Evolution of the Cosmos and Life at UCLA

In 1998, the University of California, Los Angeles (UCLA) was selected as one of the 11 initial member institutions (5 of which are universities) of the NASA Astrobiology Institute. Concurrently, in 1998-99 UCLA introduced a unique and innovative General Education cluster course called *Evolution of the Cosmos and Life*.

Evolution of the Cosmos and Life.

The course is innovative and relatively unique for several reasons. It is: (a) interdisciplinary, introducing students to both the life and physical sciences; (b) team-taught, by distinguished professors and advanced teaching fellows; (c) offered for first-year students exclusively; and (d) a year-long sequence, incorporating lectures, laboratory/discussion sections, field trips, and in the spring quarter, small satellite seminars led by the individual instructors on cluster topics of special interest to them (and the students). In its first year, 1998-99, there were four course professors, three from the Earth & Space Sciences department (geologist Jon Davidson, geochemist Mark Harrison, and paleobiologist Bill Schopf) and one from Physics & Astronomy (astronomer Matt Malkan).

Life in the Cosmos seminar
Students had two quarters of lectures and labs/discussions in the course "Evolution of the Cosmos and Life". My seminar course called "Life in the Cosmos" followed-up by looking into the variety of issues surrounding the questions of "what are the prospects for life elsewhere in the universe?" and "how do we search for evidence of such life?" The extraterrestrial life debate was looked at from different angles, including the historical and cultural perspectives as well as the current scientific approaches in astrobiology/bioastronomy.

The course was very student-centered, with lecturing minimized, and active (rather than passive) learning a key objective. The major subjects of this "Life in the Cosmos" seminar were:

1. The "Drake equation" and prospects for extraterrestrial intelligence
Critical analysis of the various factors in the formula, used to estimate the number of intelligent civilizations in the Galaxy.

2. The origin and evolution of life in the extraterrestrial context
The birth of exobiology, applying biology (and biochemistry) to considering the requirements for the origin and evolution of extraterrestrial life; exobiology/bioastronomy/astrobiology as a NASA objective.

3. Life in the solar system
Progress over the last century in our knowledge and expectations about potential life on Mars and elsewhere in the solar system, along with 21st century plans.

4. Searching for other planetary systems
Discussion of the history of this search, and current progress being made, including relatively recent discoveries of giant planets around sunlike stars and protoplanetary disks around young stars; additional emphasis on future prospects.

5. SETI
Searching for signals from extraterrestrials; the "where and how" of past and

current search programs; debating their chances of success; and their cultural/societal implications.

Some other subjects prominently discussed were (a) space exploration, travel, and colonization; challenges and prospects for our future in space; (b) \aliens and UFOs on Earth?: a history and critical study of UFO and alien abduction claims, ancient astronaut concepts, and their implications for science and society; and (c)asteroid/comet impacts: consequences for evolution, and finding (and perhaps mitigating?) potential impactors.

The two required textbooks were Dick (1998) and Jakosky (1998), which complemented each other very well, with their respective historical/cultural and scientific approaches for Life in the Cosmos. Four additional books were selected as optional/recommended (e.g. for special-topic projects): Zuckerman & Hart (1995), Goldsmith & Owen (1992), Harrison (1997), and Sagan (1997).

Course Resources

Dick, Steven J., 1998, *Life on Other Worlds: The 20th-Century Extraterrestrial Life Debate*, Cambridge University Press

Goldsmith, Donald, and Owen, Tobias, 1992, *The Search for Life in the Universe*, 2nd edition, Addison-Wesley

Harrison, Albert A., 1997, *After Contact: The Human Response to Extraterrestrial Life*, Plenum
Jakosky, Bruce, 1998, *The Search for Life on Other Planets*, Cambridge University Press

NASA Office of Space Sciences, 1999, http://www.hq.nasa.gov/office/oss/edu\-cation/edprog.htm

Sagan, Carl, 1997, *The Demon-Haunted World: Science as a Candle in the Dark*, Ballantine

UCLA AstroBiology Society, 1999, http://www.studentgroups.ucla.edu/abs/

UCLA General Education, 1999, http://www.college.ucla.edu/ge/intro.htm

UCLA GE 70 course, 1999, http://www.ess.ucla.edu/Cluster_TOC.html

UCLA Workgroup on General Education, 1997, "General Education at UCLA: A Proposal for Change," http://www.college.ucla.edu/ge/1998-1999/proposal.htm

Zuckerman, Ben, and Hart, Michael (eds.), 1995, *Extraterrestrials: Where Are They?*, 2nd edition, Cambridge University Press

Greg Schultz
University of California, Berkeley

Life in the Universe Series

Students, young and old, find the existence of extraterrestrial life one of the most intriguing of all science topics. The theme of searching for life in the universe lends itself naturally to the integration of many scientific disciplines for thematic science education. Based upon the search for extraterrestrial intelligence (SETI), the Life in the Universe (LITU) curriculum project at the SETI Institute developed a series of six teacher's guides, with ancillary materials, for use in elementary and middle school classrooms, grades 3 through 9. Lessons address topics such as the formation of planetary systems, the origin and nature of life, the rise of intelligence and culture, spectroscopy, scales of distance and size, communication and the search for extraterrestrial intelligence.

Each guide is structured to present a challenge as the students work through the lessons. The six LITU teacher's guides may be used individually or as a multi-grade curriculum for a school. The LITU project was conducted by the SETI Institute in Mountain View, CA; the project was funded by the National Science Foundation (NSF) and the National Aeronautics and Space Administration (NASA). The LITU Series is being published by Teachers Ideas Press, a division of Libraries Unlimited, Englewood, Colorado, USA. Sample lessons from each guide are available at the SETI Institute web site: http://www.seti.org. Follow the links to "Education" and then "Life in the Universe".

For Elementary School:

3. The Science Detectives (grades 3-4)
   This exciting science adventure has students trace the travels of Amelia Spacehart, an astronaut and radio astronomer who is searching the solar system for the source of a mysterious radio signal. Is the signal coming from an extraterrestrial intelligence? From her interplanetary NASA spacecraft, Amelia provides clues that lead students to explore features of the solar system, states of matter, lenses and magnification, and large scale measurements.

The SETI Academy Planet Project (Grades 5 - 6)
In this three-volume set, fifth and sixth grade students are invited to become members of the "SETI Academy." As academy members, they explore Earth's history for clues to the possible existence of life beyond our Solar System.

Designed for science classes, these volumes can be used individually or in series to create a comprehensive, thematic, scientific exploration for students.

- Evolution of a Planetary System (Volume 1)
  After exploring the evolution of our solar system, students apply that information to simulate the possible evolution of a planetary system beyond our own

- How Might Life Evolve on Other Worlds? (Volume 2)
  Through a variety of science activities, students explore the evolution of life on Earth and search for clues to the possible evolution of life on an unknown planet beyond our solar system. Synthesizing what they learn, they then learn to create life forms that could exist on that distant planet.

- The Rise of Intelligence and Culture (Volume 3)
  Students examine concepts of intelligence, culture, technology, and communication. Using their new knowledge, they simulate a complex culture for an intelligent organism that could exist on an unknown planet.

For Middle School and High School

- Life: Here? There? Elsewhere? The Search for Life on Venus and Mars (Grades 7 - 8)
  Students investigate the phenomena of life through activities that introduce them to the multidisciplinary sciences of planetology and exobiology. Simulating Venusian and Martian conditions, they explore various means of detecting life in the atmosphere and soils of Earth. They use their findings to propose a spacecraft design for life detection on Venus and Mars.

- Project Haystack: The Search for Life in the Galaxy (Grades 8 -9)
  Are there any intelligent civilizations out there? Where might they be? If they are out there, how would they communicate with us and what would they say? How would we respond? Project Haystack is an investigation of these questions and others now being asked by scientists trying to determine our place in the cosmos. Through hands-on activities, middle school students learn about the scale and structure of the Milky Way Galaxy - a cosmic haystack. They study SETI science by using simple astronomical tools to solve some of the problems of sending and receiving messages beyond our solar system.

The *Life in the Universe* Series is available from:

SETI Institute
2035 Landings Dr.

Mountain View, CA 94043
http://www.seti.org

Teacher Ideas Press
P. O. Box 6633
Englewood, CO 80155-6633
http://www.lu.com/tips/index.html
1-800-237-6124

By Edna DeVore
SETI Institute

Chapter 16 Creating Astronomy Outreach Programs

Solar Eclipse Webcasts-High Visibility Public Outreach Events

Does it seem likely that a purely educational web site on astronomy and space science could get 4 million web hits in 3 days?

That is just what happened during the Central/South America total solar eclipse webcast last February 1998. The "Live @ The Exploratorium" eclipse web site (http://www.exploratorium.edu/eclipse) received upwards of 4 million web hits, 300,000 individual sessions, and was featured in more than 50 national and international television and radio news programs (including BBC, MSMBC, CNN, etc.). The engaging natural phenomenon of a total solar eclipse was used as a "hook" to feature NASA Sun-Earth Connections science research and the scientists who investigate the effects of the active Sun on the Earth's magnetosphere. In this way, large numbers of the general public, both live at the Exploratorium science center in San Francisco, and through the Internet, were able to learn about Sun-Earth Connections.

We learned many lessons from this high-visibility public event, and decided to do a webcast from Turkey for the August 11, 1999 eclipse. Our project partnered with science centers for a variety of reasons. Partnering provided high visibility for our project and allowed us to reach large audiences. We could tap existing museum resources such as their infrastructure (including their publicity and outreach departments). They also have a tremendous ability to disseminate information to students, teachers, and the public). Museums also provide continuity and are visible contributing institutions that persist long beyond an individual project. Our role was to focus on the NASA efforts in this area, and on general space science content. We could provide the background material and the scientific expertise.

For our August 11, 1999 eclipse webcast from Turkey, we had a Girl Scout science night at Lawrence Hall of Science in Berkeley and an overnight camp-in at Exploratorium in San Francisco. We had scientists from the NASA Sun-Earth Connection project interact with Internet and live audiences. We created an archive of educational materials still available at the Exploratorium web site:

http://www.exploratorium.edu/eclipse

Our total solar eclipse event served as a draw to highlight such Sun-Earth Connection topics as solar maximum, auroras and the magnetosphere. We introduced large public audiences to NASA science with about 10 million hits and 486,000 people on web site on August 11, 1999. There were about 10,000 webcast sessions and 3500 people participated at eight museums. Media coverage was very extensive. The feedback from museum event participants was very favorable and they indicated that they felt that they learned a significant amount about the Sun. The participants particularly enjoyed the broadcast of the eclipse, the hands-on activities, the interaction with scientists, and the on-line chats with scientists.

The scientists felt that they were able to discuss their research, that the program had prepared them well, and that they would participate again in a future event. Overall, it was a very positive experience for the scientists. Future plans include a webcast of the June 21, 2001 eclipse form Southern Africa, focussing on the solar maximum theme. Webcasts of other exciting astronomical events are also being explored and we would be happy to share our experiences with other groups trying to set up such events.

Isabel Hawkins
University of California, Berkeley

Astronomy Camps

Astronomy Camps are adventures in "*doing*" science using astronomy as a teaching and discovery tool. Students become astronomers operating research telescopes, keeping nighttime hours, interacting with leading scientists, and interpreting their own observations. By operating the large telescopes (40-60 inch diameter) and electronic instruments at Mt. Lemmon Observatory (9200 feet altitude) north of Tucson, students are able to observe, record, and monitor a wide variety of celestial objects. In addition, demonstrations, field trips, and guest lectures about modern topics in space explorations help demonstrate what scientists and engineers actually do and show the importance of research in society.

Astronomy Camps are available for teenagers (7-8 days), adults (3-4 days), and educators (4 days). Costs range from $350 to $550. No prior knowledge of astronomy is required. Both beginning and advanced camps are available depending on age and mathematics background. Partial scholarships are available to teenage students based on financial need.

Professional astronomers, graduate students, and undergraduate astronomy/physics majors from around the United States serve as counselors and role models and are in contact with the students full time. The Camps began in 1988 and have now involved over 1000 students from all US states and several foreign countries. All programs are directed by me, a research astronomer and educator at Steward Observatory at the University of Arizona.

Written information, pictures, and registration materials are available from the Web site at

http://ethel.as.arizona.edu/astrocamp

Don McCarthy
University of Arizona

Science Education Gateway - A Model for Outreach

The Science Education Gateway (formerly SII) is a collaborative NASA project which brings together the expertise of NASA scientists, science museums, and K-12 educators to produce NASA science-based Earth and space science curricula for classroom and public use via the World Wide Web.

SEGway materials are produced by teachers in locally-supported collaborations with program staff at nearby partnering science museums. The partnerships support teacher-developers in achieving the goals of teaching Earth and space science online, and provide them with the training and technical support needed for their curriculum projects.

The museum partners' expertise in curriculum and Web resource development, along with their existing channels of communication with the education and scientific communities help create an active network of science education resources and users. The success of this model makes high quality Web lessons, activities, and self-guided tutorials available at no cost, providing opportunities for educators and students in many states and districts to use NASA remote sensing data.

The SEGway Web site is designed to help teachers locate and identify the resources they can use best and that fit their local curriculum and National Science Education Standards. The Web site facilitates access to SEGway materials by various user groups.

The SEGway site is at
http://cse.ssl.berkeley.edu/SEGway/

Isabel Hawkins
University of California, Berkeley

Research-Based Science Education

The Educational Outreach group at the National Optical Astronomy Observatory in Tucson, AZ, has developed a NSF-funded Teacher Enhancement program called "The Use of Astronomy in Research Based Science Education" (RBSE). This program uses data from telescopes of Kitt Peak National Observatory to engage teachers and their students in authentic research in a classroom setting. Since astronomy is primarily taught as a body of facts, students usually have misconceptions about how astronomy really is practiced. RBSE provides a way for students to actually participate in astronomy research and experience first-hand the process of doing science. Information about RBSE can be found at http://www.noao.edu/outreach/rbse. The basis concepts of RBSE, explained here, can be implemented in a variety of settings.

An effective teaching tool is "inquiry-based learning", wherein students learn through an open-ended investigation of physical phenomena. In the inquiry approach, students decide how to approach the problem, rather than be led to the answer through a systematic "cookbook" approach. The inquiry approach allows students to personally frame the question in a context best suited for them to learn meaningfully. Further, inquiry-based learning develops students' critical and analytical skills by simulating the environment in which scientific research is practiced. The goal is to get students to "do science as it is done," not just learn about it from lectures and activities far removed from the actual scientific process.

Research-based learning takes the inquiry model one step further: rather than merely simulate the research environment, students actually participate in real research. By working with real data, students get to participate in "live" science, rather than recreate old experiments which many find dull. While many students are frustrated by the many uncertainties inherent in research, most are excited by the opportunity of working on a problem of current astrophysical interest, especially a problem for which no one may know the answer. It has also been our experience that research-based learning often reaches students who are bored with traditional teaching methods and so are often labeled as "trouble students." We have also found that it is a gender friendly way to teach science concepts to female students because it lends itself to a small, collaborative group learning atmosphere as opposed to lecture-discussion.

If you are involved in research, you can develop projects which will allow your students to participate, even if only at a rudimentary level. An example of a research project on which we have students working is a search for novae in Local Group galaxies. I have data for several galaxies taken with the KPNO 0.9m telescope over the span of several years. Students "blink" these images to search for novae, which appear as stars in some epochs but are not seen in others. Another project is a spectroscopic study of AGN. Students measure the wavelengths of spectral lines, identify these lines and then determine a redshift. This information is also used to determine the identity of the object (quasar, radio galaxy, BL Lac, etc.) A good research-based learning project should give students the chance to manipulate data and analyze their results. Thus, access to computing facilities is important. Since most research projects require a year or more to complete, student goals should be set at a level that is attainable over the span of the course. Like inquiry learning the time and effort necessary is substantial, but when applied in a class of the right size and ability the reward is an outstanding means for students to learn astronomy as it is really done.

Travis A. Rector
National Optical Astronomy Observatories

Instructional Materials for Outreach: SUNSPOTS!

We have developed a lesson called *SUNSPOTS! From Ancient Cultures to Modern Times*, which can serve as a model for material developed for public outreach and suitable for broad dissemination. The Sunspots lesson is an innovative resource suitable for use in science centers and in a classroom. It has interactive Web-related activities to manipulate real-time data from NASA satellite imagery and also combines mathematics and space science curriculum elements. The Sunspots lesson was developed for grades 8-12 through partnerships between NASA scientists and educators and the Exploratorium, a well-known science museum. The Sunspots lesson covers ancient and modern solar science, safe solar observing, and sunspots' effects on space weather and on Earth. The lesson features an interactive research exercise where students attempt to correlate the areas of sunspots observed in visible light with active regions as seen in satellite X-ray imagery.

The sunspot resource is aligned with the NCTM and National Science Education Standards. It emphasizes inquiry-based methods and mathematical exercises through measurement, graphic data representation, analysis of NASA data, lastly, interpreting results and drawing conclusions. These resources have been successfully classroom tested in 4 middle schools in the San Francisco Unified School District as part of the 3-week Summer School Science curricula. This resource includes teacher-friendly lesson plans, space science background material and student worksheets. The material is packaged on a CD-ROM and printed version of the relevant classroom-ready materials and a teacher resource booklet are also available. The CD and the lesson plan resource guide are available for free by sending e-mail to outreach@ssl.berkeley.edu with a brief justification.

The lesson is also on-line at:

http://cse.ssl.berkeley.edu/segway/lessons/sunspots/

Nahide Craig
University of California, Berkeley

LodeStar Astronomy Center™

New Mexico's first astronomy-oriented science center opened its doors to the public on Tuesday, December 21, 1999. The LodeStar Astronomy Center™ is located at the New Mexico Museum of Natural History & Science in Albuquerque's Old Town. The December 21 opening was the first in a series of opening events during which new elements will be added to the 20,000 SF Center.

By the Vernal Equinox (March 21, 2000), the LodeStar Astronomy Center™ will house an advanced planetarium, an observatory with a 16" telescope, and an array of astronomy exhibits that cover "The Tools of Astronomy" and "The Search for Other Worlds."

LodeStar's™ planetarium is the world's first domed theater with high-definition, full-dome digital technology and sends visitors flying to destinations around the universe or into micro-sized worlds. The planetarium's 4,750 square foot domed screen will also display computer-generated starfields, special-effects video, and panoramic and animated slides.

"In the past year the LodeStar™ Project has reached more than 14,000 students and families in New Mexico through its travelling portable planetarium and outreach events. The LodeStar Astronomy Center™ promises to reach thousands more. "We are out to educate New Mexico students in the full range of natural sciences and the opening of the LodeStar Astronomy Center™ brings us one step closer to that goal," said LodeStar™ Executive Director John McGraw, Professor of Physics and Astronomy and Director of the Institute for Astrophysics at the University of New Mexico.

The next phase of the project is the construction of Enchanted Skies Park, a rural site devoted to an exploration the night sky. Enchanted Skies Park will be situated on a mesa top will an aerial tramway carrying visitors to an area created in the shape of our galaxy. The site will have a Dark Skies Amphitheater, a large number of telescopes for guests to use, several large telescopes that can also be operated remotely, and a research facility devoted to interferometery. Seeing at the planned site is superb, with some of the darkest, clearest, and most stable skies in North America. The Park is expected to attract visitors from throughout the world, and will have special facilities and programs for serious amateur astronomers as well.

Enchanted Skies Park will be the world's largest public park dedicated to observing and understanding the night skies and will include a forum for Native Americans to share and sustain their traditional views of the night sky.

Stephen M. Pompea
Pompea & Associates

Advice on Astronomy Public Outreach

I recently spent two years in Australia studying the public communication practices of two of that country's major observatories. Both Siding Spring and Mount Stromlo Observatories possess visitor interpretive centers adjacent to the largest telescopes on site. Such visitor centers are an observatory's best means of conducting ongoing public outreach. Visitor centers are intended to provide visitors with additional information on the site itself, and on the field of astronomy in general. The designers of the visitor centers in my study also included "education" as part of their primary mandates.

At the end of the study, I compiled several pieces of advice for anyone wishing to build an observatory that allows public access (be it a private research observatory or a publicly run community facility), and anyone wishing to construct interactive astronomy exhibitions. These bits of advice are what follow.

<u>1. Know your audience!</u>
The intended audience as defined by the exhibition designers, comprised in reality, a small segment of the entire visiting public. The result? The exhibitions were not designed for those who visit. Not only do you have to know who visits

(age groups, education backgrounds, interest levels) but also why they visit, and what they do during their visit. Only then can you design an effective center that caters to the widest possible audience. While this seems like obvious advice, it was surprising just how incompatible many features of each center were with the people using the facility. Audience analysis is crucial if you wish to conduct thorough public outreach with a visitor center.

2. Location, location, location!

By far the highlight for visitors was being allowed on the site of a research facility. One visitor, referring to the combination of observatory and on-site visitor center, quipped, "you can't have one without the other in a terribly meaningful way". Another important factor to location is the physical situation of an observatory. Most observatories are located in beautiful places, on mountaintops, far from civilization. These environments create an experience for visitors which is particularly memorable and enjoyable in itself. If you run an observatory without a significant visitor center, you could be passing up a wonderful opportunity to conduct outreach activities.

3. Context is crucial

Not everyone can build an astronomy exhibition on site at a research facility. But it is clear that creating a contextually relevant environment is of utmost importance to providing a meaningful experience for members of the public. So how do you create this context if you're not located near an observatory? The key to context is something called immersion. Immersion means including conceptually related and appropriate exhibits in a contextually relevant environment. Visitor centers at observatories are an example of a very potent immersion experience.

Another important aspect of context is "reality". Include real objects whenever possible. For example, telescopes viewing real objects in real time (everyone loves looking through a telescope). Remember that the Sun and often some planets are visible in daylight hours. Of course there's lots to see at night! Meteorites are also 'great attractors' (people love picking up a real object from outer space). Meteorites add a concrete and tangible element to a topic that many people find intangible in many ways.

Another common exhibit that has proven to be effective is one that demonstrates the weights of familiar objects on different planets. The use of things like soda cans or milk cartons again provide the concrete and relevant context for the visitor.

4. Exhibition design

Reviewing the last ten years of exhibit design research reveals that there are no hard and fast rules for designing science exhibits that are at once fun, educational, and functional. Still, amongst the various opinions there is some consensus on a few points. Most agree that an exhibit is most effective when it is hands-on, possesses meaning or is relevant to the visitor, uses concrete examples and relationships, enables the visitor to control the interaction, and is fun. Many of these ideas are also in accordance with general learning theory. While these rules apply to the design of any science exhibit intended for public consumption, the challenge to making astronomy hands on mean that these rules are important to follow from the outset of the exhibition design process. It may take some creativity but it can be done. Telescopes, meteorites and demonstrations of how much objects would weight on different planets, are all examples of exhibits that incorporate these rules.

5."Outreach" involves much more than "Education"

Most people do not visit observatories just to learn about astronomy. They visit because they're curious. They go to have a unique experience, to see something they've never seen before, to feel connected with the world and the universe in which they live. Astronomy invokes a very emotional response in people-in learning lingo. It creates an affective experience that can influence a persons attitudes, thoughts and feelings about the subject. The more emotion that is invoked during an astronomy outreach experience, the more memorable and enjoyable it becomes. Memories like this are created through experiences like looking through a telescope or picking up an actual meteorite. And again, outreach does not solely depend on having an observatory nearby. Effective outreach can be done simply by allowing people to participate through public observing programs, planetarium visits, public lectures, or visiting an actual working observatory.

6. Pretty pictures

Some physicists may frown on the idea of showing lots of "pretty pictures" to the public, for the sake of showing pretty pictures, but that's what the public loves. If pretty pictures are the door to someone's interest in the field, then give them what they want. Combining pretty pictures with sensitive and thoughtful explanations of what the picture is can provide a wonderful outreach activity and create lasting memories. Developing analogies to describe the distance scale, and size scale are important activities when interpreting astronomical images. For many people, images may be the key to their further developing support for, and a respect and interest in the field. This truly is the essence of outreach.

7. Get involved!

If you live close to an observatory or a college or university with an astronomy department, try to encourage the astronomers there (and for that matter the students) to participate in outreach activities. If you yourself are an astronomer, get involved and get your students involved! Many visitors to Siding Spring and Mount Stromlo observatories expressed interest in speaking with "real scientists", and also excitement at being allowed to see the instrument of discovery first hand.

Besides the professional community, there are also amateur astronomy organizations located pretty much everywhere. They are always happy to provide public observing events and guest speakers. If you aspire to be a public face for your field, rest assured that the greatest rewards in outreach efforts will come from the glowing faces of members of the public who so crave and delight in all things astronomical.

8. Be patient

If you do endeavor to conduct some public outreach of your own through talking to people, take to heart the following two words: be patient. Understand that they are extremely interested, but for the most part not very knowledgeable. Be prepared to get down to the basics, and use basic language. This is often a challenge in a field that is so esoteric and in many ways incomprehensible. (If the public cannot comprehend the scale of the Earth-Moon system, how can we ever expect them to comprehend the notion of a light year, or the large-scale structure of the universe?) The fun is in the trying, and you'll be appreciated for your efforts.

Kim Burtnyk
Montana State University

Roles for Scientists in Education

Scientists can play a variety of important roles in education and public outreach programs. This chart illustrates some fundamental roles. But is by no means all inclusive. The far left column constitutes various entry points into the Education and Public Outreach (E/PO) realm. The subsequent columns labeled, "ADVOCATE", "RESOURCE", and "PARTNER" represent a gradation in the time and energy spent on E/PO, with "ADVOCATE" taking the least amount of time, and "PARTNER" taking the most amount of time.

The chart was adapted from "Improving Science Education: The Role of Scientists," Bybee, Rodger W., and Cherilynn A. Morrow, Fall 1998 Newsletter of the Forum on Education of the American Physical Society

A Sampling of Roles for Scientists in Education

INVOLVEMENT LEVEL	ADVOCATE	RESOURCE	PARTNER
K-12 Students	• Participate in PTA. • Talk to school board about importance of science education.	• Judge a science fair. • Answer student e-mail. • Give tour of research facility.	• Mentor a student in your laboratory. • Partner with students in a research project.
In-Service K-12 Teachers	• Speak out in support of appropriate professional development opportunities for teachers.	• Answer teacher e-mail about science content questions. • Present in teacher workshop or some aspect of science.	• Work with a teacher to implement curriculum. • Hire a teacher intern.
Schools Of Education (Pre-Service Teachers, Graduate Students, Faculty Members)	• Speak out in your department or organization in favor of closer ties with Colleges of Education. • Promote the teaching profession in your undergraduate classes.	• Teach a science course or workshop segment for pre-service teachers. • Collaborate with education faculty to improve courses on teaching science.	• Hire a graduate in education to work as evaluator or co-developer of education project. • Develop a science course or curriculum for teachers-to-be.
Systemic Reform (District, State, National)	• Speak out at professional meetings about the importance and value of scientist involvement in systemic change.	• Review science standards for science accuracy. • Review the state framework for science education.	• Collaborate on writing or adapting science standards. • Participate on state boards for adoption of standards, instructional materials, or teacher certification.
Educational Materials Development (NSRC, EDC, Lawrence Hall)	• Speak out at a school board meeting for adopting exemplary educational materials.	• Agree to serve on an advisory board for a science education project. • Review science educational materials for science accuracy.	• Collaborate to create exemplary science education materials.
Informal Education (Science Centers, Scouts, Planetaria)	• Participate on the board of a science center, planetarium, environmental center, or museum.	• Review science content of scripts for science exhibits, planetarium shows, or environmental programs. • Give talk at a science center.	• Collaborate in creation of a museum science exhibit or planetarium show. • Serve as science coordinator for a scout troop.

For more information, go to the Space Science Institute web site education area at
http://www.spacescience.org/Education/

Cheri Morrow
Space Science Institute

Chapter 17 Astronomy Education Research

A Summary of Research about Astronomy Misconceptions

Objects and Changes in the Earth and Sky. To date, most of the educational research into student astronomy misconceptions concerning sky objects and motions has considered a heliocentric perspective. Described in the next section, enormous attention has historically been paid to explanations and models for lunar phases and the seasons. The most often noted study of astronomy misconceptions was conducted in the1980's by Sadler (1992). Often referred to as the "Project STAR" study, in reference to the NSF Instructional Materials Development project initiated following the study, Sadler found an overwhelming number of high school students could not correctly answer simple multiple-choice questions regarding the predictable motions of the day and night sky.

Schneps interviewed 23 adults standing in line at a Harvard commencement ceremony in a project that is known as the "Private Universe" video (Schneps, 1987). He found that only two of twenty-three adults could adequately explain why it is hotter in the summer than in the winter; the most cited reason was the Earth's distance from the Sun. It seems that this conception is universal. Atwood found that of 49 elementary education majors, 38 could not adequately explain seasons in a written narrative (Atwood & Atwood, 1996). Of the 49 students, 18 cited distance to the Sun as being important, 10 described the Earth's northern hemisphere being closer due to tilt, and the remaining describing a variety of reasons such as Earth's spin or the jet stream. Atwood's study went on to show that 42 of the 49 could not verbally explain seasons using models either.

Several reports in the literature confirm Sadler's results on inaccurate conceptions from a heliocentric perspective (Philips, 1991). However, there is little work describing students' conceptions of the geocentric sky. Sadler's survey-study reported both that students often think Moon phases are caused by clouds and that students do not understand the nightly motions of stars. Working independently, Reed (1972) and Chamblis (1990) found that students working with celestial spheres or in planetarium environments could substantially improve their geocentric understanding of the sky. Rollins and others found that only 79% of Texas high school students could adequately answer questions about concepts of day and night (Rollins, Denton, & Janke, 1983; Frayer, Schween-Ghatala & Klausmeier (1972).

Earth in the Solar System. According to the NSES, students in the upper elementary grades should be changing from viewing the Universe in a geocentric perspective to a heliocentric perspective. Sadler's study, although very well known, was not the first study of its kind. In 1976, Nussbaum and Novak (1976) developed a research design to investigate the progression of student notions of the Earth. It appears that student conceptions of gravity are closely tied to a conception of a spherical Earth (Vosniadou & Brewer, 1989). Treagust and Smith (1989) interviewed twenty-four grade 10 students and, from their interviews, developed a questionnaire administered to 113 students. They found that students think gravity is affected by temperature, gravity is selective about what and when it affects, and gravity is stronger at great distances in order to pull things back. Confirmed by Osborne and Gilbert (1980), they found that students believe that planets with slow or no rotation have little or no gravity and that gravity decreases with increasing distance from Sun. Philips (1991) reports that students think that gravity needs air to exist. Additional studies demonstrated that these conceptions are consistent across a variety of cultural and educational backgrounds (Nussbaum, 1979; Mali & Howe, 1979; Sneider & Pulos, 1983; Baxter, 1989).

Lunar phases are traditionally difficult for many students to understand. Using 76 elementary education majors, Callison and Wright (1993) reported that students who use physical models to explain the Sun-Earth-Moon system had significant categorical shifts from pretest to posttest but that students who were taught to only use mental models did not improve significantly. Interestingly, no significant correlation between spatial ability and model development was found. As part of a larger study, Baxter found that students often think Moon phases are caused by Earth shadows (Baxter, 1989). This idea was confirmed in independent studies by Skam (1994) and Dai (1991).

There have been few studies regarding students' conceptions of the solar system. It is generally accepted that students, and adults, often have difficulty grasping the size and layout of the solar system. As part of a larger study, Slater (1993) reported that some students think that there are hundreds of stars in the solar system, that our Sun will become a black hole, that the asteroid belt is densely populated, that the space shuttle goes to the moon each week, and that comets and meteors look the same in the sky. Vosniadou (1992) argues convincingly that students can easily accept that our Sun is hot but not that our Sun is a star because the "Sun as a star" concept is too far removed from direct experience. In fact, she reports that most students think that, of the two most prominent objects in the sky, the Earth's Moon is more like a star than the Sun is.

The Origin and Evolution of the Earth and Universe. There is an apparent lack of educational research in student conceptions or beliefs regarding these objectives. Roettger (1998) recently reported that there are some significant differences between novice and expert views of the Universe. She found that about 80% of experts she interviewed described the Universe by starting with galaxies and clusters of galaxies and reducing down to the Solar System and Earth with strong emphasis on the vast emptiness of the Universe. Students, in contrast, do not appear to have a consistent conception of the Universe and, when asked to describe the Universe, often seem to be at a loss for words. Philips (1991) reports that adults think that the Universe contains only the planets in our Solar System and that the Universe is static. Students believe the planets and Sun formed directly from the Big Bang. Lightman and Sadler (1993) report that only 25% of high school students can adequately read an HR diagram following instruction.

References

Atwood, R.K. and Atwood, V.A., 1996, Preservice elementary teachers' conceptions of the causes of seasons: Journal of Research in Science Teaching, v. 33, no. 5, p. 553-563.

Baxter, J., 1989, Children's understanding of familiar astronomical events: International Journal of Science Education, v. 11, p. 502-513.

Callison, P.L. and Wright, E.L., 1993, The effect of teaching strategies using models on preservice elementary teachers' conceptions about Earth-Sun-Moon relationships: ERIC Document ED 360 171.

Caramaza, A., McCloskey, M. and Green, B., 1981, Naive beliefs in 'sophisticated' subjects: misconceptions about trajectories of objects: Cognition v. 9, p. 117-123.

Dai, M.F.W., 1991, Identification of misconceptions about the moon held by fifth and sixth-graders in Taiwan and an application for teaching: Dissertation Abstracts International (University Microfilms No. 9124300).

Lightman, A. and Sadler, P., 1993, Teacher predictions versus actual student gains: Phys. Teach., v. 31, no. 3, p. 162-167.

Mali, G. and Howe, A., 1979, A development of earth and gravity concepts among Nepali children: Science Education, v. 63, no. 5, p. 685-691.

Morrow, C.A., 1998, Innovations in the Teaching of Astronomy: Paper presented at American Association of Physics Teachers Meeting, New Orleans, LA. AAPT Announcer, v. 27, no. 4, p. 134.

Nussbaum, J. and Novak, J., 1976, An assessment of children's concepts of the earth utilizing structured interviews: Science Education, v. 60, no. 4, p. 685-691.

Nussbaum, J., 1979, Children's conception of the earth as a cosmic body: a cross age study: Science Education, v. 63, no. 1, p. 83-93.
Osborne, R.J. and Gilbert, J.K., 1980, A method for investigating concept understanding in science: European Journal of Science Education, v. 2, p. 311-371.
Philips, W.C., 1991, Earth science misconceptions: The Science Teacher, v. 58, no. 2, p. 21-23.
Reed, G., 1972, A comparison of the effectiveness of the planetarium and the classroom chalkboard and celestial globe in the teaching of specific astronomical concepts: School Science and Mathematics, v. 72, no. 5, p. 368-374.
Roettger, E.E., 1998, Changing View of the Universe: Paper presented at American Association of Physics Teachers Meeting, New Orleans, LA., AAPT Announcer, v. 27, no. 4, p. 4.
Rollins, M.M., Dentton, J.J. and Janke, D.L., 1983, Attainment of selected earth science concepts by Texas high school seniors: The Journal of Educational Research, v. 77, no. 2, p. 81-88.
Sadler, P., 1997, Students' Astronomical Conceptions and How They Change: Paper presented at American Association of Physics Teachers Meeting, Phoenix, AZ. AAPT Announcer, v. 26, no. 4, p. 78.
Sadler, P., 1992, The initial knowledge state of high school astronomy students: Ed.D. Dissertation, Harvard School of Education.
Schneps, M.H., 1987, A Private Universe: Available from Pyramid Films and Video, 2801 Colorado Avenue, Santa Monica, CA 90404.
Skam, K., 1994, Determining misconceptions about astronomy: Australian Science Teachers Journal, v. 40, no. 3, p. 63-67.
Slater, T.F., 1993, The effectiveness of a constructivist epistemological approach to the astronomy education of elementary and middle level in-service teachers: Ph.D. Dissertation, University of South Carolina.
Sneider, C. and Pulos, S., 1983, Children's cosmographies: understanding the earth's shape and gravity: Science Education, v. 67, no. 2, p. 205-221.
Treagust, D.F. and Smith, C.L., 1989, Secondary students understanding of gravity and the motions of planets: School Science and Mathematics, v. 89, no. 5, p. 380-391.
Vosniadou, S., 1992, Designing curricula for conceptual restructuring: lessons from the study of knowledge acquisition in astronomy: ERIC Document ED 404 098.
Vosniadou, S. and Brewer, W.F., 1989, The concept of Earth's shape: a study of conceptual change in childhood: ERIC Document ED 320 756.
Zeilik, M. Schau, C., Mattern, N., Hall, S., Teague, K.W., and Bisard, W., 1997, Conceptual astronomy: a novel model for teaching postsecondary science courses: Am. J. Phys., v. 65, no. 10, p. 987-996.

Jeff Adams and Tim Slater
Montana State University

Multiple Intelligences

Science educators should be aware of the theory of multiple intelligences . This theory was formulated by Harvard psychologist Howard Gardner and was discussed in detail in his 1983 book *Frames of Mind*. The theory of multiple intelligences serves to broaden the concept of human intelligence. It argues that intelligence is much more than what is measured from the narrow perspective of an IQ test.

Gardner described in his book the human capacity for solving diverse kinds of problems. He also sought to describe the human potential for creativity in a context-rich, natural setting. He grouped human capabilities into seven categories, or "intelligences." It is the challenge of the good teacher to take into account as many of these intelligences as possible when teaching or designing curriculum and materials.

Linguistic Intelligence: The capacity to use words, either orally or in writing. People with high linguistic intelligence include poets, storytellers, orators, journalists, and joke writers.

Logical-Mathematical Intelligence: The capacity to use numbers, and to use logic. This intelligence is used to categorize, classify, infer, generalize, calculate, and to test hypotheses.

Spatial Intelligence: This form of intelligence involves a sensitivity to line, shape, color, form, space, and spatial relationships. This intelligence comprises a visualization capacity, and the ability to orient oneself spatially under a wide variety of spatial circumstances. Architects, inventors, builders, mechanical engineers, and interior decorators, are examples of people who probably possess good spatial intelligence.

Bodily-Kinesthetic Intelligence: This intelligence centers on the ability to use your body to express yourself. Possessors of this form of intelligence have physical skills like coordination, dexterity, balance, and flexibility. People gifted in this area might include mechanics, woodworkers, sculptors, surgeons, dancers, and athletes.

Musical Intelligence: This capacity includes a heightened awareness of and ability to discern pitch, rhythm, melody, tone color, and other musical attributes.

Interpersonal Intelligence: This includes the ability to be sensitive to and make distinctions in the feelings, moods, and intentions of others. Sensitivities to the voice, facial expressions, gestures and body language of others is part of this intelligence, as well as a capacity to respond to these cues.

Intrapersonal Intelligence: This intelligence deals with self-knowledge and the ability to have a realistic and accurate self-perception. This includes knowing one's strengths and limitations, moods, and intentions. High self-esteem and self-understanding are indicators of this form of intelligence.

How can you appeal to students with these differing abilities to the greatest extent possible, given you teaching style? For me, the main practical point that I take from the theory of multiple intelligences is that each student must be valued and appreciated as an intelligent individual. Even if your class is skewed towards a severe appreciation of Logical-Mathematical Intelligence, have some humility in the sense of realizing that this is not the only form.

If a student does not perform well on the kinds of tasks you value, your job is to assist them in increasing their skill without putting the student down for their "lack of intelligence". In a different culture, or at a different time, the people smart in Logical Mathematical intelligence might be wearing the dunce caps, because they cannot perform very well with some other form that is highly valued. We should all strive to develop skills and greater abilities in all of the above areas. I have had the experience many times of former students who were very average in my class come back and tell me about their great successes and graduate degrees. When this happens I can appreciate better that these individuals have a great many more skills than can be exhibited in a traditional science education classroom.

For more information on how the idea of multiple intelligences can be applied in the classroom and in developing curriculum materials, see the book *Multiple Intelligences in the Classroom*, by Thomas Armstrong, published by the Association for Supervision and Curriculum Development (ASCD). The ASCD web site has links to articles on multiple intelligences as well.

http://www.ascd.org/

Stephen M. Pompea
Pompea & Associates

Alternative Assessment in Astronomy

The purpose of alternative assessment strategies is to evaluate the actual process of doing science or mathematics instead of only grading the product or final result. The most common of these is a performance assessments which can be designed to provide the student with either a single, holistic "level of competency rating" or a series of ratings over several aspects of a task. The key to effective and meaningful performance tasks is the use of authentic tasks. An authentic task requires students to address real-life problems that simulate problems researchers face. In contrast to traditional problems, these tasks are characterized by being complex, somewhat undefined, engaging problems that involve sustained work and require the effective utilization of text or computer resources. When used with students with highly varying abilities, performance tasks can take maximum advantage of multiple entry and exit points and emphasize multiple-correct answers and approaches.

Students are graded on the process of problem solving as well as the product using a rating scale based on explicit standards. The goal of a performance assessment is to judge the level of competency for the process of doing science, mathematics, and research. Accordingly, performance assessment strategies are best used to monitor student process skills and problem solving approaches. The most effective graded tasks are authentic tasks that are open-ended with multiple-correct solutions when the instructor can observe the student. Here is an example:

Telescope Performance Assessment

Your task is to set up and align the 8″ telescope, find three different sky objects, and accurately describe some interesting aspects of these objects.

Level 3: Student completes all aspects of task quickly and efficiently and is able to answer questions about the equipment used and objects observed beyond what is obvious.

Level 2: Student completes all aspects of task and is provides descriptive information about the equipment and objects observed.

Level 1: Student is not able to complete all aspects of task or is not able to sufficient provide information about the equipment used or objects observed.

Level 0: No attempt or meaningful effort obvious.

Tim Slater
Montana State University

Constructivism

If you have a room full of science teachers at different grade levels and you need to sort them into secondary school teachers and college teachers, how do you do it? It's very easy. All you have to do is ask how many know what "constructivism" is? Constructivism is an important topic that secondary science and mathematics teachers are talking about extensively. College science teachers usually have not heard of the term.

Constructivism is not a theory about teaching, but rather a theory about learning and how knowledge is created. You can read about the research in constructivism at hundreds of web sites. Here are several, for example.

"Constructivism and the Five E's", at the Miami Museum of Science website
http://www.miamisci.org/ph/lpintro5e.html

University of Colorado at Denver School of Education
http://carbon.cudenver.edu/~mryder/itc_data/constructivism.html

Constructivism, Technology, and the Future of Classroom Learning
http://www.ilt.columbia.edu/k12/livetext/docs/construct.html

Stephen M. Pompea
Pompea & Associates

I. Using Collaborative Groups

For two years, we have been successfully utilizing small learning groups to keep students engaged in the course content during class. This technique often goes by the broader names of "collaborative learning" or "cooperative groups" but the idea is basically the same. We challenge groups of students working together to solve a problem, complete a task, or create a product. The approach is based on the idea that learning is a social activity in which the students talk among themselves. It is through these social interactions that learning occurs. An extensive research base on collaborative learning already exists and we would like to describe how we are using collaborative groups to support learning in our large-lecture course for non-science majors with the hope that readers will be able to modify this technique for successful implementation at their own institutions.

To more actively engage our students, we developed a series of 30-minute duration group activities in four categories: historical, conceptual, process-oriented, and product oriented. These activities were designed to maximize interactions among students by focusing on open-ended questions that engender student discussion and avoiding tedious calculations. For example, the first activity challenges students to use a telescope catalog and decide how best to spend $6000 to outfit a community astronomy program. Using the vocabulary of education, such an activity has multiple-entry and exit points and the task can be approached at varying levels of complexity based on the student population.

As another example, we provide students with 16 photographs of galaxies and ask them to devise and defend a classification system for the collection. Certainly some students naturally adopt the same tuning-fork system that Edwin Hubble advocated, but many others are also creatively presented. Without exception, the activities that work best for collaborative groups are ones in which there are multiple correct solutions. When we do use activities where there is only one correct solution—using parallax to determine the distance to an object for example—the level of student engagement seems to drop substantially. The activities do, however, require significant structure to support groups working independently as a single instructor cannot answer all the questions that might arise in a completely open-inquiry format.

Successfully implementing student learning groups requires a level of "buy in" from the students. We explain to the students, both in the lecture and in the syllabus, that we are using collaborative learning groups to make them more actively involved in their own learning and that we believe that they will learn more from one another than they will from our lectures. We also tell them that we love to lecture, but that we believe that allocating class time to this activity is worthwhile. Interestingly, it seems that students need to be reminded occasionally why they are working in collaborative groups, so we do so periodically. Students are told that their group will turn in just one activity and they will all share the same grade for their work. In our class, activities constitute about 25% of the course grade, the remainder being carried by homework, quizzes, and exams. Students are probably most surprised when we tell them that they will work in their groups during examinations. During exams, we give each student two answer sheets. They first work on the exam independently and submit an individual answer sheet. Then, they collaborate with their group and complete the same exam a second time. After discussing their answers, they submit a second answer sheet. This second answer sheet allows them to disagree with their group if they so choose. In the end, the first (individual) and second (group discussion) exam scores are averaged. Scores are naturally higher the second time, but this allows us to ask more difficult questions without the students getting too irate.

Jeff Adams and Tim Slater
Montana State University

II. Organizing Collaborative Groups

The most inexact part of collaborative group approaches involves how the groups are composed. Formally assigned grouping is a formidable task in the large lecture course from a management perspective. The most widely adopted approach is that student groups should be diverse in backgrounds, ideas, ethnicity, and gender. The common sense idea is that science-friendly students will support and help teach the science-phobic students. However, a growing body of research is beginning to contradict this idea of heterogeneity. This research suggests that isolating students of color from other students of color or women from other women can impede the academic success of these individuals because they can become isolated, distracted, or placed in stereotypical roles even in the small group environment.

In our large classes, we have the students self-form groups. We tell our students that it is important to be in a group of people who share their level of interest and class attendance style. The rare group meltdowns usually seem to occur either when one member of the group wants to go slow and understand every part of the learning activity when other members want to get to a final answer and finish as soon as possible. Or they may occur when fellow group members repeatedly fail to attend class on activity days. Still, we find that most groups are formed by people who are accidentally sitting close together or who know each other socially. We allow students to change groups but this rarely occurs. Despite these potential problems, most students readily report that they enjoy in-class collaborative learning groups.

Jeff Adams and Tim Slater
Montana State University

III. Student Roles in Groups and Further Tips

Group roles are used. On each activity task sheet, we list four blanks for student names next to the four roles we ask students to fulfill; the roles rotate with each new activity. The task of the *leader* is to be sure that each member of the group contributes, that everyone's ideas are represented, and that the group stays on task to finish the activity in the allotted time. The role of the *recorder* is to write the group consensus answers on the answer sheet and to be sure that the assignment is turned in before the class is finished. The *skeptic's* role is to ask the question, "are we sure?" and "why do you think that?" The *explorer's* task is to try to investigate ideas and areas that no one else has considered. For the occasions that there are only three members of a group present, we suggest that students forego the explorer's role. Through focus group interviews, we have found that students do not always adhere to these roles with the exception of the recorder. Accordingly, we remind students that the reason roles are used is because the class time is so limited, they do not have an opportunity for the natural group roles of their members to emerge and rotating roles allows everyone to participate.

Initially, we aggressively enforce a rule of four students to a group. This is because when a student is absent, then there are still three members present. If a group has only two members, then the likelihood that they need help from the instructor seems to go up; something worth avoiding in a classroom of 200 students. Serendipitously, one gift revealed itself in grading student assignments. Instead of 200 papers to grade, there are only 50 papers to grade when there are four students per group.

A few more lessons learned are worth describing. The first is about the student seating arrangement. Our lecture hall is composed of long tables running from one side of the room to the other. When students attempted to work by sitting four-across, invariably, one or two people were always left out. We strongly urge students to sit two in one row and two in the next in the shape of a square. One early idea we use to encourage students work together was to provide only one set of instructions for each group.

In this approach, we have not abandoned lecture entirely; we estimate that 70% of the class time is still allocated to lectures. However, by focusing on engendering an environment where students are active learners, we have made significant improvements in our course reputation. Enrollments are steadily growing as students learn about the course. Just as important, we are having fun teaching the course even if we do not get to lecture the entire time. We interact individually with students much more often than we did before as we spend our time going from group to group asking probing questions. In the end, we spend more time working directly with students and, in turn, students spend more time actively engaged in their learning.

Jeff Adams and Tim Slater
Montana State University

A Berkeley Compendium of Suggestions for Teaching with Excellence

This web site present a series of suggestions for improving teaching. This compendium describes more than 200 teaching techniques that faculty members have found to be effective in their courses at the University of California, Berkeley. These suggestions cover the most of the major aspects of university teaching ranging from planning and preparing courses to presenting material. It also covers motivating students and the giving and getting feedback on learning.

It is a valuable reference tool to help teachers their own teaching. It is more of a resource site than a narrative meant to be read in one sitting. It is recommended that you begin by reading the initial menu and skimming the Index in order to locate those sections or particular suggestions of greatest interest to you. The compendium was compiled by Barbara Gross Davis, Lynn Wood, and Robert C. Wilson.

http://uga.berkeley.edu/sled/compendium/

Stephen M. Pompea
Pompea & Associates

Astronomy Diagnostic Test (ADT) Information

The development of this diagnostic test was accomplished the multi-institutional Collaboration for Astronomy Education Research (CAER) including, among many others, Jeff Adams, Rebecca Lindell Adrian, Christine Brick, Gina Brissenden, Grace Deming, Beth Hufnagel, Tim Slater, and Michael Zeilik.

The test is free for introductory astronomy faculty members to use, with some agreed guidelines. The test is not to be altered in any way, questions are not to be left out, questions are not to be removed and placed on course quizzes or exams, and the test items are not to be distributed to students lest the items find their way into "student files." In exchange for using the ADT, the authors respectfully request that you submit scores to the national database (please e-mail Beth Hufnagel for details). The complete test can be found in the appendix.

Tim Slater
Montana State University

Astronomy Education Research Annotated Bibliography

From Tim Slater, Montana State University

J.P. Adams and T.F. Slater, "Using Action Research to Bring the Large Lecture Course Down to Size," Journal of College Science Teaching, 28(2), p. 87-90 (1998). **A series of astronomy action research studies including a high correlation between student self-report knowledge and examination scores.**

J.P. Adams and T.F. Slater, "Astronomy in the National Science Education Standards," Journal of Geoscience Education, 48(1), 39-45 (2000). **Literature review of astronomy education research organized by the NSES content learning objectives.**

R. K. Atwood and V. A. Atwood, "Preservice elementary teachers' conceptions of the causes of seasons," Journal of Research in Science Teaching, 33(5), 553-563

(1996). **The authors propose that misconceptions about seasons might not be as firmly entrenched as commonly thought.**

J. Baxter, "Children's understanding of familiar astronomical events," International Journal of Science Education, 11, 502-513 (1989). **Many student conceptions are consistent across cultures.**

P. L. Callison and E. L. Wright, "The effect of teaching strategies using models on preservice elementary teachers' conceptions about Earth-Sun-Moon relationships." ERIC Document ED 360 171 (1993). **Students who use physical models during instruction have larger achievement gains than students who only use mental models.**

A. Caramaza, M. McCloskey, and B. Green, "Naive beliefs in 'sophisticated' subjects: Misconceptions about trajectories of objects," Cognition 9, 117-123 (1981). **Students can often give correct answers to test questions without having accurate mental models.**

C. R. Chamblis, "A planetarium oriented sequence of exercises," In The Teaching of Astronomy, J. M. Pasachoff and J. R. Percy, (Eds.). (Cambridge University Press, Cambridge, U.K. 1990), pp. 387-388. **The manipulation of celestial spheres and visits to interactive planetarium presentations improve student understanding of sky motion.**

M. F. W. Dai, "Identification of misconceptions about the moon held by fifth and sixth-graders in Taiwan and an application for teaching." Dissertation Abstracts International (University Microfilms No. 9124300), (1991). **Students think that moon phases are caused by Earth's shadow on Moon.**

J.E. DeLaughter, S. Stein, C.A. Stein, and K.R. Bain, "Preconceptions abound among students in an introductory earth science course," AGU EOS Transactions, 79(36), 429-432 (1998). **Student-supplied-response exam shows students have difficulty describing a spherical Earth with a flat surface, that students have many contradictory conceptions, and that planetary motions are not understood. Data and test available on the Web at**
http://www.earth.nwu.edu/people/seth/Test

F. M. Goldberg and L. C. McDermott, "Student difficulties in understanding image formation by a plane mirror," Phys. Teach. 24(8), 472-481 (1986) and F. A. Goldberg and L. C. McDermott, "An investigation of student understanding of the real image formed by a converging lens or concave mirror," Am. J. Phys., 55, 108-119 (1987). **Exploration of student difficulties with prediction strategies and semantics that physicists use when describing images, mirrors, and lenses.**

D. Hestenes, M. Wells, and G. Swackhammer, "Force Concept Inventory," Phys. Teach. 30(3) 141-158 (1992) and I. A. Halloun and D. Hestenes, "The initial state of college physics students," Am. J. Phys. 53, 1043-1056 (1985). **Describes the rigorous development a nationally adopted conceptual test in physics education that is used to determine the effectiveness of various instructional techniques for students' conceptions of Newton's three laws of motion.**

Lightman and P. Sadler, "Teacher predictions versus actual student gains," Phys. Teach., 31 (3), 162-167 (1993). **Teachers predict much higher student achievement gains than students can actually achieve.**

G. Mali, and A. Howe, "A development of earth and gravity concepts among Nepali children," Science Education, 63(5), 685-691 (1979). **Students' conceptions of gravity are highly dependent on the specifics of their mental model of a spherical Earth.**

National Research Council, National Science Education Standards (National Academy of Sciences Press, Washington, 1996). **Describes a complete framework vision for a national content, instruction, assessment, program, and teacher professional development in science education. Available on the Web at http://www.nap.edu/books/readingroom/nses/**

J. Nussbaum, "Children's conception of the earth as a cosmic body: a cross age study," Science Education, 63(1), 83-93 (1979). **Students have a wide variety of mental models of Earth as a sphere.**

J. Nussbaum, and J. Novak, "An assessment of children's concepts of the earth utilizing structured interviews," Science Education, 60(4), 685-691 (1976). **Describes a highly effective and systematic interviewing strategy appropriate for investigating student conceptions.**

R. J. Osborne and J.K. Gilbert, "A method for investigating concept understanding in science," European Journal of Science Education, 2, 311-371 (1980). **Describes a highly effective and systematic interviewing strategy appropriate for investigating student conceptions finding students have misconceptions about gravity.**

W. C. Philips, "Earth science misconceptions," The Science Teacher, 58(2), 21-23 (1991). **A long list of student conceptions in earth and space science based on classroom experience of the authors.**

G. Reed, "A comparison of the effectiveness of the planetarium and the classroom chalkboard and celestial globe in the teaching of specific astronomical concepts," School Science and Mathematics, 72(5), 368-374 (1972). **Students learn sky motions equally as well with a celestial sphere as in an interactive live planetarium teaching classroom.**

E. R. Roettger, "Changing View of the Universe." Paper presented at American Association of Physics Teachers Meeting, New Orleans, LA. AAPT Announcer, 27(4), 4 (1998). **Professional astronomers usually describe the universe as being very large and full of mostly empty space often starting with Earth and moving to larger entities. Alternatively, students usually don't have a holistic view of the Universe or are at ease to describe it.**

M. M. Rollins, J. J. Dentton, and D. L. Janke, "Attainment of selected earth science concepts by Texas high school seniors." The Journal of Educational Research, 77(2), 81-88 (1983) using methods described by D.A. Frayer, E. Schween-Ghatala, and H. J. Klausmeier, "Levels of concept mastery: implications for instruction," Educational Technology, 12(12), 23-29 (1972). **Less than 80% of Texas high school students can accurately describe the cause of day and night.**

P. Sadler, "The initial knowledge state of high school astronomy students," dissertation, Harvard School of Education Dissertation Abstracts International, 53(05), 1470A. (University Microfilms No. AAC-9228416)(1992). **A comprehensive survey of high school student astronomy knowledge.**

Sadler, P. M. Psychometric Models of student conceptions in science: Reconciling qualitative studies and distractor-driven assessment instruments. Journal of Research in Science Teaching, 35, 265-296 (1998) and P. Sadler, "Students' Astronomical Conceptions and How They Change," Paper presented at American Association of Physics Teachers Meeting, Phoenix, AZ. AAPT Announcer, 26(4), 78 (1997). **Students can move from one misconception to another misconception on their journey to scientifically accurate conceptions. Use of Item Response Theory.**

M. H. Schneps, A Private Universe (1987 videotape). Available from Pyramid Films and Video, 2801 Colorado Avenue, Santa Monica, CA 90404 (1987). **Video of Harvard graduates and middle school students describing seasons and moon phases using faulty mental models.**

P. S. Shaffer and L. C. McDermott, "Research as a guide for curriculum development: an example from introductory electricity." Am. J. Phys. 60(11), 994-1013 (1992). **One of a series of seminal papers from physics education describing how educational research is used to create research-based curriculum materials.**

C. Sneider and S. Pulos, "Children's cosmographies: understanding the earth's shape and gravity," Science Education, 67(2), 205-221 (1983). **Students think gravity is determined by proximity to Sun and rotation rate of planets.**

K. Skam, "Determining misconceptions about astronomy," Australian Science Teachers Journal, 40(3), 63-67 (1994). **Students believe that moon phases are caused by Earth's shadow.**

T. F. Slater, "The effectiveness of a constructivist epistemological approach to the astronomy education of elementary and middle level in-service teachers," Ph.D. Dissertation, University of South Carolina (1993). **The same effective approaches used with children work well with in-service teachers to build confidence as well as knowledge. Extensive description of what constructivism looks like in astronomy.**

T.F. Slater, J.R. Carpenter, and J.L. Safko, "A constructivist approach to astronomy for elementary school teachers, " Journal of Geoscience Education, 44(5) (1996). **Using hands-on activities founded in constructivism, elementary teachers significantly increased their astronomy knowledge and attitudes toward teaching it.**

T.F. Slater, J.L. Safko, and J.R. Carpenter, "Long Term Sustainability of Teacher Attitudes in Astronomy." Journal of Geoscience Education. 47, p. 366-368, 1999. **Teachers who learned astronomy in a specially designed course maintained positive attitudes, values, and interests four years after the experience.**

D. F. Treagust and C. L. Smith, "Secondary students understanding of gravity and the motions of planets," School Science and Mathematics, 89(5), 380-391 (1989). **Students think gravity is determined by proximity to Sun and rotation rate of planets.**

S. Vosniadou, "Designing curricula for conceptual restructuring: lessons from the study of knowledge acquisition in astronomy." ERIC Document ED 404 098 (1992). **Students can easily accept that our Sun is hot but not that our Sun is a star because the "Sun as a star" concept is too far removed from direct experience. Similarly for day and night.**

S. Vosniadou and W. F. Brewer, "The concept of Earth's shape: a study of conceptual change in childhood." ERIC Document ED 320 756 (1989). **Student's conceptions of a spherical Earth are highly distorted.**

M. Zeilik, C. Schau, N. Mattern, S. Hall, K. W. Teague, and W. Bisard, "Conceptual astronomy: a novel model for teaching postsecondary science courses," Am. J. Phys., 65(10), 987-996 (1997). **Undergraduate courses using collaborative groups and concept maps erase many ethnic and gender differences that exist in pre-test scores.**

Chapter 18
Astronomy Teaching and the Science Education Standards

Astronomy in the NRC National Science Education Standards

A review of the NSES content strands reveals numerous places where astronomy and space science concepts are either explicitly or implicitly described as an important part of a student's science education. Explicitly, in the earth and space sciences content strand, the K-12 astronomy objectives can be grossly organized, with sincere apologies to the conscientious NSES authors, into eleven age-appropriate conceptual objectives, which are shown in Figure 1. A discussion of the equally important unifying concepts and processes in science, science as inquiry, relationships to life and physical science, science and technology, nature of scientific knowledge, and the social perspective of science is beyond the scope of this manuscript.

Grades K-4 Objectives. From the perspective of age-appropriate instruction, geometrical and abstract astronomy in the early grades is indeed problematic. The NSES suggest that the goals of astronomy education at the K-4 level should focus on describing the properties, locations, and motions of sky objects from a geocentric perspective. These sky objects include the Sun, Moon, stars, clouds, birds, and airplanes. Some parents might believe that a child who can recite "101 fun facts" about the planets is impressive, but the NSES focuses on the underlying processes and themes of science instead of facts. This is not to imply that memorized factual knowledge is not important—constellation names can be, and should be, learned in the same way as the names of farm animals, the multiplication tables, and the months of the year. However, knowing names of constellations and the order of the planets at the knowledge level is only the first step in an astronomical knowledge base and therefore must not form the core of the learning objectives. Instead, as has been done with NSES, the objectives must be framed so students can be expected to fully participate in the process of science at an age-appropriate level. For example, at the K-4 level, it is appropriate for students to observe and chart the changing phases of our Moon. To memorize the phase cycle would disengage students from the scientific process and be wholly inappropriate; it would be equally inappropriate though for K-4 students to be expected to create mental models of the abstract geometry of the Sun-Earth-Moon system. Likewise, seasons should be studied from a geocentric perspective with an emphasis on student observations.

Grades 5-8 Objectives. In the culture of US education, the last formal astronomy for the vast majority of students is in grades 8 or 9 in the context of an earth science curriculum. It is at this level that the NSES expects students to describe the solar system from a heliocentric (Sun-centered) perspective. At this level, the standards propose that students should be using the heliocentric motion of the solar system to explain the phenomena of day/night, seasons, eclipses, and lunar phases. At this stage, students should be comparing and contrasting the solar system objects in terms of terrestrial/non-terrestrial, satellite systems, rotation/revolution, size/mass, and characteristics that students find important.

The NSES suggest that the concept of gravity should be introduced at the 5-8 level. Students should have an accurate model of a spherical Earth where gravity holds people to Earth's surface and explains tides. Through a heliocentric study of solar system motions, students should understand that gravity causes the planets to orbit in nearly circular orbits as described by Kepler's Laws. This conceptual foundation of gravity is revisited in secondary level objectives when describing the characteristics of galaxies and star clusters

Grades 9-12 Objectives. The four secondary-level astronomy objectives focus on the observation, origin, evolution, and characteristics of the Universe beyond the solar system. Possibly even controversial in some schools, the NSES call for students to understand current scientific explanations of the origin and evolution of the Earth (multiple atmospheres), the Solar System (nebular hypothesis), the elements (nucleosynthesis), and the Universe (the Big Bang). Comprehension of the observational and theoretical evidence for exotic objects such as neutron stars and black holes is indeed abstract; yet, these are the same topics that provide cover stories for popular newsmagazines. In many ways, today's students are more familiar with these news stories than basic naked-eye astronomy. Interestingly, except for the very small percentage of students taking high school physics or, even rarer, astronomy, in grades 11 or 12, there currently appears to be little astronomy taught in US schools at the secondary level. The NSES provide a needed impetus for increasing the quantity of astronomy included in the secondary curriculum.

Astronomy Concepts in the National Science Education Standards
Sky objects have properties, locations, and movements that can be observed and described. [K-4]
The sun provides the light and heat necessary to maintain the temperature of the earth. [K-4]
Objects in the sky have patterns of movement. The sun, for example, appears to move across the sky in the same way every day, but its path changes slowly over the seasons. [K-4]
The earth is the third planet from the sun in a system that includes the moon, the sun, eight other planets and their moons, and smaller objects, such as asteroids and comets. [5-8]
Most objects in the solar systems are in regular and predictable motion. those motions explain such phenomena as the day, the year, phases of the moon, and eclipses. [5-8]
Gravity is the force that keeps planets in orbit around the sun and governs the rest of the motion in the solar system. Gravity alone holds us to the earth's surface and explains the phenomena of the tides. [5-8]
The sun is the major source of energy for phenomena on the earth's surface, such as growth of plants, winds, ocean currents, and the water cycle. Seasons result from variations in the amount of Sun's energy hitting the surface, due to the tilt of the earth's rotation on its axis and the length of the day. [5-8]
The sun, the earth, and the rest of the solar system formed from a nebular cloud of dust and gas 4.6 billion years ago. The early earth was very different from the planet we live on today. [9-12]
The origin of the universe remains one of the greatest questions in science. The "big bang" theory places the origin between 10 and 20 billion years ago, when the universe began in a hot dense state; according to this theory, the universe has been expanding ever since. [9-12]
Early in the history of the universe, matter, primarily the light atoms hydrogen and helium, clumped together. Billions of galaxies, each of which is a gravitationally bound cluster of billions of stars, now form most of the visible mass in the universe. [9-12]
Stars produce energy from nuclear reactions, primarily the fusion of hydrogen to form helium. These and other processes in stars have led to the formation of all the other elements. [9-12]

The National Science Education Standards (NSES) are available on-line at:
http://www.nap.edu/readingroom/books/nses/html/

Jeff Adams and Tim Slater
Montana State University

Misconceptions about the National Science Education Standards

Here are seven misconceptions scientists often have about the National Science Education Standards (NSES).

Misconception 1: The National Research Council Science Education Standards are required for all schools.

When a person hears the word "standard" they tend to infer the nuance of "requirement". Moreover, "national standard" tends to imply a requirement for the nation. Nevertheless, the NRC Standards are optional, and state education systems interpret them and relate to them in many different ways, some more embracing than others. For example, many states have developed a set of state science content standards that derive directly from the guidelines provided by national documents such as the NRC standards and the AAAS Benchmarks. On the other hand, California is a notable example of a state whose science education standards have deviated significantly from the NRC standards. In addition, states have varying requirements on local school districts to conform to their state standards. As a result, one must investigate a given school district's or school's relationship to national standards on a case-by-case basis.

The educational system in the US is in stark contrast to most other countries that do not have standards per se, but instead have required national curricula. It is important to remember that Standards do not provide curriculum, and so in America it is up to curriculum developers in partnership with scientists and teachers nationwide to voluntarily design courses of instruction and student experiences that lead to learning the fundamentals articulated in the Standards.

Misconception 2: The National Science Education Standards are a list of scientific facts students should know.

To many people, the term "education standard" naturally translates to "fact a student should know". The NRC standards are much more than a collection of facts. The *content* standards alone describe what all students – regardless of background or circumstance – should understand and be able to do at different grade levels from kindergarten through high school. The NRC content standards are differentiated by grade level (K-4, 5-8, and 9-12).

The NRC content standards are organized under the following headings: Unifying Concepts and Processes in Science, Science as Inquiry, Physical Science, Life Science, Earth and Space Science, Science and Technology, Science in Personal and Social Perspective, and History and Nature of Science. Note that perspectives and experiences with the way science works and evolves are at least as heavily emphasized as the fundamental ideas and concepts in science.

A very common misconception is that the Standards involve content areas only, but the NRC document provides guidelines for systemic change that go well beyond this. The document also includes standards for teaching, student assessment, the professional development of teachers, science programs in schools and school districts, and for science education policies in school systems. For example, the NRC teaching standards advocate an "inquiry-based" approach to classroom instruction that has many similarities with the way modern scientists *practice* science.

Misconception 3: Standards-based methods of science teaching are just the latest fad and we should return to the basics that worked so well for me in school.

Most of today's scientists, including those in our workshops as well as those who engaged in our project to create an educator guide, have been taught science (and thus continue to teach science) in traditional ways, involving predominantly lectures, memorization, and textbooks.

Misconception 4: "Standards-based" means "hands-on activities"

A standards-based lesson offers the educator a sound approach to instruction based on the best available research about how students learn and what teaching practices facilitate that learning. This often involves the use of what are commonly called "hands-on" activities, but as may be discerned from Table 1 above, such activities by themselves are insufficient to make the lesson "standards-based". In addition, Table 2 summarizes the difference between a conventional approach to teaching, a hands-on approach, and an inquiry-based approach aligned with Standards. The differences are illustrated in the context of teaching a biology lesson about trees. The examples clearly show that "hands-on" is a necessary but not sufficient quality for being "inquiry-based", and so it is possible for a lesson to be "hands-on" without being aligned with Standards.

Misconception 5: The NRC Standards apply only to curricular materials.

No, the NRC Standards address many other educational activities besides the development of curricular materials. For example, there are standards that articulate best practices in how to professionally develop teachers. An example of a conventional approach to a teacher "workshop" would be an all-day series of lectures by scientists.

By contrast, teacher workshops that are aligned with standards are themselves models for inquiry-based instruction (see Tables 1 and 2). Such professional development opportunities provide participants with appropriately tailored background and experience, both with the pertinent science and with the best instructional practices. They provide direct hands-on experience with Standards-based lessons that teachers can readily use in the classroom (such curricular materials are best disseminated in conjunction with teacher training). Standards-based professional development also follows up with workshop participants to support them in their efforts to improve their teaching practice.

Misconception 6: It is easy to link my scientific research topic to Standards and thus make a K-12 educational product or activity that is "standards-based" or "aligned with Standards".

It is often easy to make general intellectual links between a scientific research topic and many of the content Standards, but this is not the same thing as aligning with Standards. Aligning an educational product or activity with the NRC Standards is a challenging, multi-dimensional task that is very often underestimated.

For example, to align a lesson with the Standards requires awareness of the research about how students learn, what misconceptions they commonly have, how their capabilities change with age, and what instructional and assessment methods work best – all of this in addition to a deep understanding of the fundamental science being taught by the lesson. Successful standards-based curricular materials are also subject to review, field testing, and revision before being broadly disseminated. The development process takes significant time, money, and expertise in science, in classroom teaching, and in standards-based curriculum design.

For example, NASA's Cassini mission will study the Saturn system. There are no science education standards that say students should learn all about the research conducted by the Cassini mission. However there *are* Earth and Space science education standards that call for the study of the Solar System and the planets. Standards also say students should learn about Systems, Order, and Organization, about Science as a Human Endeavor, and about the relationship between technology and scientific discovery. Cassini's exploration of the Saturn system (as well as other research projects) can provide a motivational context for such standards-based learning.

Misconception 7: Practicing teachers generally know all about the Standards and how to apply them

Not necessarily so. Generally, good K-12 teachers are experts in managing a classroom of active young people. This is another multi-dimensional challenge, often underestimated by scientists and others those who have not tried it on a day-to-day basis. Teachers also know what types of activities really engage their students' curiosity, and they recognize the attributes of lesson plans and educator guides that are the most accessible, easy-to-use, and effective in the classroom setting.

Although teachers are generally expert at adapting all sorts of ideas for use in their classrooms, they may not necessarily be trained or experienced in *consciously* applying the Standards. In our educator guide project, three of the four exceptional classroom teachers with whom we worked had never perused a copy of the NRC Standards, let alone prepared standards-based curricular materials for use by other teachers. Of course, the scientists in our partnership started out even less familiar with the Standards, and they too learned a great deal throughout our development process. The scientists' gift was a touchstone to real-world exploration and a deep understanding of the process and content of science.

The challenging task of designing curriculum aligned with Standards required us to add team members and reviewers who were experienced in applying and evaluating principles of standards-based curriculum design. We augmented our development team of teachers and scientists with a curriculum design consultant and an evaluator, both from the local university's college of education. We were fortunate to have a curriculum specialist who had also spent considerable time in the classroom. Such a hybrid professional proved to be very valuable in helping to bridge the gaps of understanding and experience between scientists, teachers, and educational specialists. She worked with us to include direct quotes from the Standards in our educator guide to help elucidate the intended learning goals for teachers. Our workshops for scientists include presentations and discussions led by such education specialists as well as by practicing teachers.

Cheri Morrow
Space Science Institute

Comparing Approaches to Teaching and Assessment

The table below emphasizes the differences between traditional and inquiry-based learning, and amplifies further the similarities between the practice of science and the practice of science teaching. Note that "hands-on" is a necessary but not sufficient quality for being "inquiry-based".

Comparing Approaches to Teaching and Assessment
(developed in collaboration with Tim Slater)

Conventional Approach	Hands-On Approach	Inquiry-Based Approach
The teacher tells students that trees can be classified by examining their bark and their leaves. She shows pictures of trees in a textbook and asks students to memorize the names of the different types of trees according to the sort of bark and leaves they have.	The teacher tells students that trees can be classified by examining their bark and their leaves. She shows pictures of trees in a textbook and takes students to the park and asks them to match the pictures with the real trees. She asks students to memorize the tree names for the test. The class moves on to another hands-on activity about plants and flowers.	The teacher tells students that scientists classify trees by the different features they have. She asks them to come up with ideas for what features would distinguish one tree from another. She takes them to the park to explore their ideas and to make observations and gather data that would help them create their own classification scheme for trees. She asks them to compare to established classification schemes and to present reports on their results. She follows up with a lesson about the nature and classification of trees in other climates.

Note that an exemplary inquiry-based lesson sets the stage for a new lesson that builds on the old one. Ideally, inquiry-based teaching does not consist of a string of isolated, "one-shot" activities anymore than science consists of a string of isolated "one shot" experiments. In both science and reformed science teaching the results of one experiment (or activity) answers some questions and raises others that can be addressed by new experiments.

For more information, go to the Space Science Institute website education area at http://www.spacescience.org/Education/

Cheri Morrow
Space Science Institute

Scientific Research Approach vs. Inquiry-Based Teaching Approach

Scientists often fail to recognize that "the basics that worked so well" for them in school may not work so well for the majority of other students who will not become scientists. An extremely effective way of illustrating potential problems with traditional teaching and assessment methods was provided by one of our presenters at the1995 Space Science Institute workshop for scientists on K-12 education. Professor Jay Hackett of the University of Northern Colorado showed the following viewgraph:

The Monotillation of Traxoline
(attributed to Judy Lanier)

It is very important that you learn about traxoline. Traxoline is a new form of zionter. It is monotilled in Ceristanna. The Ceristannians gristerlate large amounts of fevon and then bracter it to quasel traxoline. Traxoline may well be one of our most lukized snezlaus in the future because of our zionter lescelidge.

Directions: Answer the following questions in complete sentences. Be sure to use your best handwriting.

1. What is traxoline?
2. Where is traxoline monotilled?
3. How is traxoline quaselled?
4. Why is it important to know about traxoline?

The presentation of the "Traxoline" story illustrates how it is possible to "pass the test" without really knowing or understanding much of anything. Students can memorize and use sophisticated vocabulary words, but be devoid of any deeper conceptual understanding rooted in their own experience. It remains for many scientists to discover and internalize that modern science education reform is about teaching students in a so-called "inquiry-based" fashion – a way that bears enormous similarity to how scientists *practice* science as opposed to how they learned it in school. The table below offers an analogy between the approach to scientific research and an approach to teaching science aligned with Standards. It has shown some early promise in assisting scientists with conceptual change about science teaching.

Scientific Research Approach	Inquiry-Based Teaching Approach
Raise fundamental question of interest that is addressable via scientific investigation.	Engage student interest; guide the development of questions [i.e. establish basis for **inquiry**) in a specific area of content.
Research what is already known	Discuss with students what they already "know" or think they know **[prior knowledge assessment]** to help address the question(s).
Make a prediction or hypothesis in answer to the question of interest	Ask students to make a prediction or hypothesis in answer to the question of interest
Plan and implement an experiment to test the prediction.	Plan and implement an experiment to test the prediction **[hands-on activity]**
Reflect on the results of the experiment and how they affect what was known before. Be alert for how the new data does or does not readily fit into the existing structure of scientific understanding.	Reflect with students on the results of their hands-on activity/investigation and use their predictions to assist them with gaining new/deeper understanding of content. Be alert for any shifts from "prior knowledge" as students integrate their new experiences.
Communicate new knowledge via talks and papers. Science community judges the validity and value of the results. New questions are raised.	Communicate new knowledge via presentations, papers, demonstrations, exams **[assessment methods]**. Teachers judge students' learning and guide them to apply it to new circumstances.

For more information, go to the Space Science Institute website education area at http://www.spacescience.org/Education/

Cheri Morrow
Space Science Institute

Chapter 19
A Resource Guide for Astronomy Teaching

Introduction

Resources for astronomy teaching are abundant. This resource guide can only be a teaser, since there is not enough room here to include even a few percent of the resources available. In general, I have picked my favorite resources gathered from 25 years of astronomy teaching. I am sure that many great books and other high quality resources have been left of the list. Even though this list reflects my tastes and interests I hope that it proves useful. If you have a resource that you think should be on the list please send me information about it. My thanks to Drs. Karen Meech, Elizabeth Roettger, and Tim Slater for providing material for inclusion in this area.

If you take away only one thing from this resource area, it should be that the Astronomical Society of the Pacific is the astronomy teacher's best friend.

> A note on web addresses.
> Web addresses change almost as fast as moon phases. The web addresses are provided as a convenience. I have included enough information with each resource so that a search using a decent search engine and search word will lead you to the right place most of the time. The easiest way of searching is often to put quotation marks around a few words likely to appear on the web site.

The Astronomical Society of the Pacific

The Astronomical Society of the Pacific (see http://www.aspsky.org/) is the starting point for filling up on astronomical teaching resources. They have a terrific print and on-line catalog of astronomy education resources. Andrew Fraknoi, former executive director of the Society had done a tremendous amount of work compiling teaching resources.

The result of his efforts in this area is a resource notebook entitled *The Universe at your Fingertips*, created through the support of the Banbury Fund and distributed by the A.S.P. This is probably the most comprehensive collection ever assembled of materials and resource lists for the teaching of astronomy. If you are serious about astronomy teaching, join the A.S.P. and connect yourself to a large network of astronomy educators and to a wealth of well-tested curriculum and supplemental materials.

The Astronomical Society of the Pacific also has a terrific on-line astronomy education resource edited by Andrew Fraknoi. It includes information on Project ASTRO, on astronomy education in the US, and a listing of national astronomy education projects. It also includes an astronomy education bibliography, information on women in astronomy, and resources on astronomical pseudo-science. There are extensive discussion and resources on light pollution and environmental issues and astronomy. Teachers will find especially useful the column "Black Holes to Blackboards" from Mercury magazine written by master high school and college astronomy teacher Dr. Jeff Lockwood and available on this web site.

http://www.aspsky.org/education/educ_bib.html

As I mentioned the *Universe at Your Fingertips* notebook is a key product of Project ASTRO. This loose-leaf notebook contains: exemplary classroom activities selected by a team of teachers and astronomers, comprehensive resource lists and bibliographies, brief background material on astronomical topics, and teaching ideas from experienced astronomy educators. The notebook also includes a catalog of national astronomy education projects. A follow-on companion volume entitled *More Universe At Your Fingertips* edited by Dennis Schatz of the Pacific Science Center should also be available as well by the time this book goes to press. Both *Universe at Your Fingertips* books are available from:

Astronomical Society of the Pacific (ASP)
Catalog Department
390 Ashton Avenue
San Francisco, CA 94112 USA
Toll Free in the U.S. (800) 335-2624

Andy Fraknoi has also put together a list of resources for teachers of introductory astronomy which includes:

- Collections of Course Syllabi
- Introductory Astronomy Textbook Web Sites
- Suggestions for Better Teaching Approaches
- Demonstrations for the Classroom
- Laboratory and Observing Exercises
- Applets and Other Web-based Exercises
- Interdisciplinary Approaches to Astronomy Teaching
- Miscellaneous Resources

This extremely valuable reference is available is available on the ASP web site: http://www.aspsky.org/education/educsites.html

Basic Reference Books

Encyclopedia of the Solar System
Edited by Paul R. Weissman, Lucy-Ann McFadden, Torrence V. Johnson, Academic Press, 794 pages, 1998
Comprehensive recent reference on all things planetary including a chapter on searching for extra-solar planets.

Encyclopedia of Astronomy and Astrophysics
Edited by Robert A. Meyers, Academic Press, 1989
One of the best encyclopedias about astronomy. Very well illustrated source of in-depth knowledge on a variety of subjects

The Cambridge Atlas of Astronomy, 3rd edition
Edited by Jean Audouze, Jean-Claude Falque, Guy Israel, Cambridge University Press, 1994
A useful, well-illustrated, oversize volume.

Handbook of Space Astronomy and Astrophysics, Second Edition
Martin V. Zombeck, Cambridge University Press, 1990
One of the most valuable reference guides available. Roughly equivalent to Allen's *Astrophysical Quantities* with more emphasis on space astronomy.

The Planetary Scientist's Companion
Katharina Lodders, Bruce, Jr. Fegley, Oxford University Press, 1998
Recent, comprehensive reference on all aspects of planetary science. An essential extension to both Zombeck's and Allen's book. Full of tables on all aspects of the major and minor solar system bodies.

Glossary of Astronomy and Astrophysics, Second Edition
Jeanne Hopkins, The University of Chicago Press, 1980
The best astronomical dictionary available. Very clear and precise.

Hubble Atlas of Galaxies
Allan Sandage, Carnegie Institution of Washington, 1984
A classic book that brings pleasure to many by its elegance.

Modern Physics
Kenneth S. Krane, John Wiley & Sons, 1995
My favorite modern physics book. Includes relativity, quantum physics, and applications.

Observational Astronomy Books

Nightwatch
Terence Dickinson, Spiral paperback edition, Firefly Books, 1998
If I could only pick one book to give to a person with an interest in the night sky, this would be it. Actually, I have given away many copies over the years and many of my relatives and friends have benefited from this excellent book.

Star Maps for Beginners
I. M. Levitt and Roy K. Marshall, Simon and Schuster, 1992.
The best book of star maps and detailed mythological stories. The star maps are the easiest set to use, bar none. The book has a planetary positions table, but they may be out of date if you buy an older edition. However, the star maps and stories will always be fine, so buy any copy of this book you can find.

Skyguide, Revised edition
Mark R. Chartrand, Golden Press, 1990
A very useful pocket guide. Suitable for use as a text in observational astronomy courses. One of the very best books with detailed information on each constellation. Very useful for star party discussions.

Observer's Handbook
Editor: Roy Bishop, The Royal Astronomical Society of Canada
Highly useful guidebook published every year. Full of information on every aspect of observational astronomy in an easily accessible format. Includes a teaching guide by Professor John Percy, a well-respected astronomy educator.

A Field Guide to the Stars and Planets (Field Guide to the Stars and Planets, 4th Ed)
by Jay M. Pasachoff, Houghton-Mifflin 1999. Indispensable guidebook for teachers. Very well written. A classic book for the anyone interested in the sky.

Astronomical Tables of the Sun, Moon, and Planets, 2nd edition
Jean Meeus, Willman-Bell, 1995
A very useful book to find out the exact times of celestial phenomena several decades in advance. For example, the opposition times of Mars are given for the years 0-3000AD.

National Audubon Society Field Guide to the Night Sky (Audubon Society Field Guide Series)
Mark R. Chartrand, Knopf, 1991.
Useful field guide with monthly star charts.

StarList 2000
Richard Dibon-Smith, Wiley Science, 1992.
A handy star catalog that describes the characteristics of over 2,000 stars to a visual brightness of 5.25. Very useful compendium of data about stars, arranged by constellation. A good field guide for the serious amateur.

The Moon Book: Fascinating Facts About the Magnificent, Mysterious Moon
Kim Long, Johnson Books, 1998. A handy guidebook to all aspects of the moon's movement and surface. Also includes sections on eclipses and photographing the Moon. Easy to read and useful for students.

Star Names Their Lore and Their Meaning, Revised edition
Richard Hinckley Allen, Dover Publications, 1963
A classic book on the history of star names in various cultures. The book is filled with many ancient and classical literary references that will appeal to students of history.

Colours of the Stars
David Malin and Paul Murdin, Cambridge University Press, 1984
Beautifully illustrated. Discusses many of the basic astronomy and detection principles behind the pretty pictures we see on posters. A superb book, which covers almost every astronomical topic.

The Mapping of the Heavens
Peter Whitfield, Pomegranate Artbooks, 1995.
Lovely book on the history of star maps and charts around the world. Lavishly illustrated.

Stairways to the Stars : Skywatching in Three Great Ancient Cultures
Anthony Aveni, John Wiley & Sons, 1999
A detailed examination of how the Incans, Mayans, and the people who built Stonehenge made sense of the heavens. Written by a professor of astronomy and anthropology at Colgate University.

The Sky: A User's Guide
David H. Levy, Cambridge University Press, 1994
Excellent introduction to all aspects of observational astronomy. Extensive discussion of planetary observing. Includes a section on poetry about the night sky.

The Ever-Changing Sky : A Guide to the Celestial Sphere
James B. Kaler, Cambridge University Press,1996.
Probably the best reference book on time, orbital motions, movements in the sky, sunrise and sunset times, and celestial navigation. Well-written and contains answers to many questions not addressed in standard college astronomy textbooks. Highly recommended.

The Invisible Universe Revealed: The Story of Radio Astronomy
Gerrit L. Verschuur, Springer-Verlag 1987
Well-written book on a popular level about radio astronomy.

Skywatching (Nature Company Guide)
David H. Levy, John O'Byrne (Editor)
Time Life; 1995
Good introduction to observational astronomy.

Books on Optics and Observing Techniques

Astronomical Observations
Gordon Walker, Cambridge University Press, 1987
Excellent introduction to observing techniques, telescopes, and instrumentation. Includes material on the construction of sensitive low noise detectors, the preservation of image quality. Very broad in its topics.

Astrophysical Techniques, 3rd edition
C. R. Kichin, Inst of Physics Pub, 1998
Very good discussions of detectors, imaging, and photometry.

Light and Its Uses
Readings from the Amateur Scientist column of Scientific American, W. H. Freeman, 1980.
Full of high-level home projects for experimentally minded students. Sample projects include building spectrographs and photometers. Perfect for independent study projects.

Telescope Optics : Evaluation and Design
by Harrie G.J. Rutten, Martin A.M. Van Venrooij, 1988, Willmann-Bell

Excellent, highly technical reference on telescope optics

Choosing and Using a CCD Camera
Richard Berry, Wilman-Bell, Inc. 1992. Excellent book on CCDs and image processing. Well written, understandable to students and well-illustrated.

Optics, 3rd edition
Eugene Hecht, Addison-Wesley, 1997
My favorite introductory optics book. Well-illustrated and full of insights.

Alice and the Space Telescope
Malcolm Longair, Johns Hopkins University Press, 1989
Excellent book on the HST and the science it can do. The story is told along the lines of Alice in Wonderland.

Books on Demonstrations and Experiments

Building Scientific Apparatus, subtitled A Practical Guide to Design and Construction, 2nd edition
John H. Moore, Christopher C. Davis, and Michael A. Coplan, Christopher C. Davis, 1989 Perseus Press
An excellent reference book on all aspects of scientific equipment. From glassblowing to detectors and lasers. The emphasis is on practical and useful information. Indispensable for experiments.

Turning the World Inside Out and 174 Other Simple Physics Demonstrations
Robert Ehrlich, Paperback - 216 pages,1990 , Princeton University Press

Why Toast Lands Jelly-Side Down: Zen and the Art of Physics Demonstrations
by Robert Ehrlich Paperback, 224 pages,1997 , Princeton University Press

What Light Through Yonder Window Breaks?: More Experiments in Atmospheric Physics
by Craig F. Bohren Wiley Science Editions, Paperback - 190 pages,1991, John Wiley & Sons
Very nice book on atmospheric optics.

The Flying Circus of Physics
Jearl Walker, John Wiley and Sons, 1977.
Great book on physics of the "everyday" world. Contains many simple and profound demonstrations. Many seminars have been based on this book. Using this book is a great way to get students to investigate nature.

Books about the Process of Doing Science

What Do You Care What Other People Think?' : Further Adventures of a Curious Character
Richard Feynman, Bantam Doubleday Dell Pub, 1992
This is the Feynman book that recounts the Challenger investigation. I find myself rereading it from time to time and I always get something more out of it.

Chalk Up Another One, The Best of Sydney Harris, 1992, and *Einstein Simplified: Cartoons on Science*, 1989, both from Rutgers University Press.
Books of cartoons with poignant scientific commentary. Great for starting off lectures.

The Man Who Sold the Milky Way, subtitled "A Biography of Bart Bok"
David H. Levy, University of Arizona Press. 1995.
This book by itself provides more evidence than one could hope for about how human an activity astronomy is. A well written, fun-to-read biography of a very influential astronomer.

In Quest of Telescopes
Martin Cohen, Cambridge University Press 1980.
A classic astronomical travelogue. A personal account of observing and the life of an astronomer.

Clyde Tombaugh, Discoverer of Planet Pluto
David Levy, University of Arizona Press, 1991
Great biography that shows that joy and passion of the Tombaugh

Technology of Our Times, People and Innovation in Optics and Optoelectronics
Edited by Frederic Su, SPIE Press, 1990
Twenty-five chapters about leading scientists and engineers and how they worked. Each leader tells a short story about his/her work and life. Very lively, exciting, and characteristic of the ups and downs of a life in science and technology. Many of the greatest experiments are described in personal detail.

Disturbing the Universe
Freeman Dyson, Harper and Row, 1979
Reflective book on the life of Freeman Dyson, a brilliant physicist. Many stories about the scientists he worked with and the creative projects with which he was involved. Dyson addresses the role of science in society through a variety of stories. This is the kind of book one rereads regularly over many years. The chapter on Dyson's work with the British Bomber Command during WWII is particularly poignant. Highly recommended.

Books on Solar Astronomy

Solar Astronomy Handbook
Rainer Beck, Heinz Hilbrecht, Klaus Reinsch, and Peter Volker, Willmann-Bell, 1995.
A fine book on solar telescopes, solar filters, special instruments, photography, and all varieties of solar observations. A useful reference for amateurs and professionals, and a good guide for eclipse observers.

Sun, Earth, and Sky
Kenneth R. Lang, Springer-Verlag, 1997
An outstanding, well-illustrated book on the Sun and solar-terrestrial phenomena. Good discussions of the Sun's impact on life on Earth.

Fire of Life, The Smithsonian Book of the Sun, Smithsonian Books, 1981.
One of the most accessible books on the Sun. Traces history of attitudes about the Sun, sunpower, the dynamic Sun, and our place in the cosmos.

The Great Sundial Cutout Book
Robert Adzema and Mablen Jones
A great book for teachers of the earth and space science. Using this book students can produce accurate
sundials and gain an understanding of how they work.

Books on Life in the Universe

Earth : Evolution of a Habitable World
Jonathan Lunine, Cambridge Univ Press, 1999
Great overview of the history of the Earthwritten from a planetary perspective. Great introductory textbook for
earth science or astronomy coursess.

The Search for Life on Other Planets
Bruce Jakosky, Cambridge University Press,1998
Excellent book that addresses the question "Does life exist on other planets?"

Extraterrestrials : Where Are They? 2nd edtiion
Ben Zuckerman (Editor), Michael H. Hart (Editor), Cambridge University Press, 1995.
The book consists of 22 chapters written by scientists and others who examine the plausibility of the existence of other intelligent life forms. A good book for a seminar course.

Sharing the Universe : Perspectives on Extraterrestrial Life
Seth Shostak, Berkeley Hills Books; 1998
A well-written book which describes what the scientific community believes extraterrestrial "intelligent" beings could be like. Includes chapters on searching the solar system for life, extending the search to the stars, alien abilities and behavior, motives, Drake Equation, and SETI. Addresses the reaction of society if E.T. should call.

Intelligent Life in the Universe
Carl Sagan, I.S. Shklovskii, Emerson-Adams Press, 1998
A classic book, still of great value in teaching about this topic.

What if the Moon Didn't Exist? Voyages to Earths That Might Have Been
Neil F. Comins, HarperCollins, 1993.
A thought-provoking book that explores ten hypothetical situations about our Earth and solar system. Valuable for students who take the world for granted. Very well written.

Books on Astronomy Education or Activity Books

Astro Adventures
Dennis Schatz and Doug Cooper, Pacific Center, 1994
Well-loved and well-used book of astronomy activities. General topics include Moon Gazing, Sun Watching, Star Finding, Planet Picking, and Further Adventures. Highly recommended.

Astronomy Education : Current Developments, Future Coordination
Edited by John R. Percy (Astronomical Society of the Pacific Conference Series Proceedings ; Vol 89, 1996
Proceedings of an Astronomical Society of the Pacific Symposium held in College Park, MD in June, 1995.
Discussion of many of the astronomy education projects and development. A valuable resource.

New Trends in Astronomy Teaching
Edited by L. Gouguenheim, D. McNally, and J. R. Percy, Cambridge University Press, 1998.
Papers presented at IAU Colloquium 162 held at University College London and the Open University, July 8-12 1996
Excellent resource on worldwide astronomy education projects, experiments, and developments.

Planetarium Activities for Student Success, 3rd edition
Series Editors: Cary Sneider, Alan Friedman, and Alan Gould. Published jointly by Lawrence Hall of Science, U. California, Berkeley and the New York Hall of Science.
A series of 12 books about effectively using a planetarium. Some of the books present complete planetarium programs with related classroom activities. Others are full of resources. The philosophy of the book series is that planetarium programs can be highly participatory.

Interactive Lesson guide For Astronomy, Cooperative Lesson Guide for Astronomy
Michael Zeilik, , The Learning Zone, Inc. , Sante Fe, New Mexico, 1998
A very useful guide for implementing cooperative learning in the classroom. Covers nearly every astronomy subject matter area. An excellent resource to "jump start" your implementation of cooperative learning.

Mysteries of the Sky: Activities for Collaborative Groups
Jeff Adams and Tim Slater, Kendall/Hunt Publishing Company, 1998
A great resource for introducing collaborative learning into the classroom. Very effective way to increase interest and participation in observational astronomy.

Books on Teaching Topics and Techniques

Just-In-Time Teaching : Blending Active Learning With Web Technology
Gregor M. Novak, Evelyn T. Patterson, Andrew D. Garvin, and Wolfgang Christian
Prentice Hall Series in Educational Innovation,
Just-in-Time Teaching is an new teaching and learning methodology that engages students using feedback from pre-class web assignments to adjust classroom lessons. Students receive a rapid response to specific questions and problems rather than just hearing a generic lecture. The process also encourages students to take more responsibility for their learning. This resource book for educators provides tested resource materials that can be used in introductory physics classrooms.

Peer Instruction: A User's Manual
Eric Mazur, Paperback, 253 pages, Prentice Hall1996.
Contains a large variety of class-tested, ready-to use resources that emphasize teaching by questioning. Understanding is stressed above memorization, and the book contains many questions designed to assess conceptual understanding. One chapter is titled "A Step-by-Step Guide for a Peer Instruction Lecture". The book includes two resource floppy disks.

Science Teaching Reconsidered : A Handbook
by Committee on Undergraduate Science Education
National Research Council/National Academy Press, 1997.
Excellent resource handbook on new developing solid science teaching at the college level. Includes details on specific methods of teaching, evaluation and grading, choosing and using instructional materials, and getting to know your students. A most useful introduction to many of the most solid, useful trends in science education.

Teaching About Evolution and the Nature of Science
National Academy of Science, National Academy Press, 1998
Available on-line and for order from www.nap.edu
Superb book on every aspect of teaching about evolution. Very useful for the astronomy, biology, or earth science teacher at high school or college level. Also of great value for an administrator or interested member of the public. Has a variety of excellent student activities and good advice on how to deal with controversy surrounding the teaching of evolution at all levels. Written by a team of experts who have probably been dealing with this issue all of their professional lives.

They're Not Dumb, They're Different : Stalking the Second Tier
Sheila Tobias, Paperback, 94 pages 1994 ,Science News Books
This book is a must read for teachers of introductory college course. Tobias argues that science must open it doors to the "second tier" students who have decided not to pursue a career in science for various reasons. These students are not teacher-proof and curriculum-proof and are sensitive to how the course is taught and the atmosphere of the course. These students require more support and attention. If you believe that scientists are born, not made, perhaps you should read this book.

Colorado State University Teaching Resources Manual
Written and compiled by Frank J. Vattano
Distributed by The Office of Instructional Services, The Graduate School, CSU, Fort Collins, Colorado, 1998.
Excellent resource manual designed for graduate teaching assistants. (Very useful for college science teachers of all ages). If you are a university professor and your school does not have such a manual, you should get this one. Dr. Vattano is a superb teacher (also a Professor of Psychology and a former Dean as well) and has co-taught for many years a course for graduate students on college teaching. (Why doesn't every university has such a professional preparation and development course for future faculty members?) I took his course in 1977; I am still talking about it!

Other Books

Cosmos
Carl Sagan, Ballantine Book, 1993
A classic and well-worth having

Cosmic Clouds : Birth, Death, and Recycling in the Galaxy
James B. Kaler, W H Freeman & Co 1997
Well-written, well-illustrated book on star formation.

Fire in the Sky: Comets and Meteors, the Decisive Cneturies in British Art and Science
Roberta J. M. Olson and Jay M. Pasachoff, Cambridge University Press, 1998. A chronicle of the development of British visual art concerning comets and meteors from the late Renaissance to our own century.

Magazines

Astronomy (Kalmbach Publishing)
Giffith Observer (From the Griffith Observatory, Los Angeles)
Mercury (from the Astronomical Society of the Pacific)
New Scientist (from the U.K.)
Planetary Report (from the Planetary Society)
Sky and Telescope (Sky Publishing)
StarDate (from the University of Texas McDonald Observatory)

Slides, Videotapes, and Videodiscs

The Astronomical Society of the Pacific (see above for contact information) has a good selection of slides and videos. Your local or regional planetarium is also a good source. NASA's Central Operation for Resources for Educators (CORE) is a very good sources for videos as well.

The ASP slide sets include:

- Grand Tour of the Universe
- Planetary Nebulae
- The Search for Planets around Other Stars
- Star Maps and Constellations
- Hubble Space Telescope images (7 sets)
- The Earth from Space
- Moon
- Secrets of the Sun

- Ancient Life on Mars
- Worlds in Comparison
- Galileo Mission to Jupiter
- Teacher's Solar System Slide set
- The Jupiter Impact

NASA CORE Catalog

The NASA Central Operation of Resources for Educators (CORE) catalog has more than 200 NASA produced videocassette, slide, and CD-ROM programs available for a minimal charge. This Adobe Acrobat PDF version is provided for your convenience to download and view the CORE products off-line. This is a great resource for teachers! It has an amazing array of material.
http://spacelink.nasa.gov

Slides and Videos at the Center for Astrophysics

Slide Sets
The CfA does not offer for sale individual slides or photographs of general astronomical objects or its facilities. The following three slide collections are available for purchase. For details on acquiring slides, please call (617) 495-7461 or email pubaffairs@cfa.harvard.edu
-Colors of the Universe
-The Electronic Sky
-Visions of Einstein

Video on Annie Jump Canon
From Dr. Debra Elmegreen:
"Readers of the AAS Women's Network electronic newsletter may be interested to learn of a recently completed, 25-minute educational video now available from the Harvard-Smithsonian Center for Astrophysics. "Annie and the Stars of Many Colors: A Portrait of Astronomer Annie Jump Cannon, 1863-1941" explores the life and work of Annie Cannon through the eyes of sixth graders who visit the Harvard College Observatory, where Cannon worked for nearly 50 years and developed her important stellar classification scheme. Conceived, written, and narrated by CfA historian of science Barbara Welther, "Annie and the Stars of Many Colors..." was created to inspire young students, especially girls and minorities, to learn about astronomy, astronomers, women in science, and history in general.

By following the students as they visit archives, participate in hands on science activities, and talk with CfA astronomers, "Annie and the Stars of Many Colors..." shows young people actively engaged in the process of discovery, and offers teachers, guidance counselors, and others a novel tool to stimulate discussion about science, history, women's studies, and career choices. An accompanying teacher's guide with additional resource information is now being developed.

The video was produced using staff, resources, and facilities at the CfA and its Science Media Group. Funding for the project was provided by the Smithsonian Institution's Women's Committee and the Educational Outreach Fund of the Institution's Office of Education and External Affairs. For more information on purchasing"Annie and the Stars of Many Colors," please call (617) 495-7461 or email pubaffairs@cfa.harvard.edu.

Astronomy Phonelines with Recorded Messages

Abrams Planetarium (East Lansing, Michigan) 517.332.7827
Flandrau Science Center Planetarium & Observatory, University of Arizona, Astronomy Hotline: 520.621.4310
Hansen Planetarium (Salt Lake City, Utah) 801.532.7827
Sky and Telescope Skyline (Boston, Massachusetts) 617.497.4168

For more detailed astronomy related information intended for both beginning and advanced amateur astronomers
nationwide go to:

Sky and Telescope magazine's "What's up (in the Sky)?"
http://www.skypub.com/whatsup/whatsup.html

Abrams Planetarium Sky Calendar
http://www.pa.msu.edu/abrams/diary.html

Pacific Science Center Astronomy Resources
http://www.pacsci.org/public/exhibits_planetarium/astronomy/astrodefault.html

Resources on Women in Astronomy

Women of NASA
Includes some bilingual resources and Chats
http://quest.arc.nasa.gov/women/

Four Thousand Years of Women in Science
Includes introductions, biographies, references, and photographs. Has an extesive list of related we b sites.
http://crux.astr.ua.edu/4000WS/4000WS.html

Committee on the Status of Women in Astronomy "Women in Astronomy" Web site
The site can be used to
--find speakers for colloquia, meetings, or schools
--solicit job applicants
--sort by education, expertise, research interests, etc. for statistical or search purposes
http://www.stsci.edu/stsci/service/cswa/women/

Women in Astronomy: An Introductory Bibliography
by Andrew Fraknoi and Ruth Freitag
http://www.aspsky.org/education/womenast_bib.html

The Woman Astronomer is published quarterly and is about promoting astronomy and describing the women role-models in astronomy, past and present. It is also about the quickly changing science, and hobby, of astronomy.
http://users.erols.com/njastro/twa/

Expect the Best from a Girl Web site
From the Women's College Consortium
Describes what parents can do at home and at school to encourage their daughter to pursue interests in math and science. Has links to resources and describes several women role models. Useful information about women and work and a sampling of programs for girls.
http://www.academic.org/index.html

Women Nobel Prize Laureates
Has short biographies of women laureates and some book references, with links to women in science and women in technology pages. The main site is at
http://nobelprizes.com/nobel/nobel.html
while the pages devoted to women laureates are at:
http://www.almaz.com/nobel/women.html

Bilingual Education Resource Material

Universo Online home page. Spanish Language astronomy resource
http://universo.utexas.edu/

Resources: Bilingual Education Sites
http://www.phoenix.k12.or.us/pes/cuarto/bauer/bisites.html

Resources: Bilingual Education Materials
http://www.phoenix.k12.or.us/pes/cuarto/bauer/bimaterials.html

Some Useful Web Pages in Astronomy

SEGway,
Science Education Gateway (SEGway) is a great place to get free curriculum material developed for K-12 on a variety of astronomy subject matter areas. SEGway is a NASA sponsored project hosted by the Center for Science Education at the University of California, Berkeley Space Sciences Lab.
http://cse.ssl.berkeley.edu/SEGway/

NASA SpaceLink
The starting point and best way to tap into NASA's extensive resources for educators
http://spacelink.nasa.gov

The Exploratorium
This hands-on science center web site is a great resource for new events and demonstrations.
http://www.exploratorium.edu/

Solar Max 2000
A terrific resource for understanding the solar maximum. Should we be concerned? Find out at this web site. The site has current solar images and events.
http://www.exploratorium.edu/solarmax/index.html

Hubble Space Telescope Images
http://oposite.stsci.edu/pubinfo/pictures.html

The Astronomy Picture of the Day
http://antwrp.gsfc.nasa.gov/apod/astropix.html

The National Space Science Data Center (NSSDC) Photo Gallery
Extensive image gallery
http://nssdc.gsfc.nasa.gov/photo_gallery/

Welcome to the Planets
Planetary imagery from JPL
http://pds.jpl.nasa.gov/planets/

Ideas and References for Demos in Astronomy, from Dr. Caty Garmany of the University of Colorado
Nice site with many good links
http://lyra.colorado.edu/sbo/katy/seminar/demos.html

Home page of the "Windows to the Universe" Project at the University of Michigan.
A terrific astronomy education resource project.
http://www.windows.umich.edu/

Extrasolar Planet Search Project at San Francisco State University
http://www.exoplanets.com/

Extrasolar Planets Encyclopedia
This website at the Observatoire de Paris is a comprehensive encyclopedia of all aspects of extrasolar planet research.
http://www.obspm.fr/encycl/encycl.html

University of Michigan's Observational Mishaps page
Shows what can go wrong when you are observing
http://www.astro.lsa.umich.edu/users/kaspar/obs_mishaps/mishaps.html

Satellites and Satellite Tracker Page at NASA's Marshall Space Flight center (general home page is http://liftoff.msfc.nasa.gov/). Good resource for tracking satellites.
http://liftoff.msfc.nasa.gov/academy/rocket_sci/satellites/

The University of Oregon's Astronomy HyperText Book
Many valuable resources and applets.
http://zebu.uoregon.edu/text.html

AstroEd: Astronomy Education Resources, a HPCC/ESS K-12 Outreach Project at the University of Washington. Nice list of curriculum materials and on-line courses
http://www-hpcc.astro.washington.edu:80/scied/astro/astroindex.html

International Planetarium Society. A great resource for all things domed.
http://www.ips-planetarium.org/

The Observatorium
Great source for earth and space science information. Nice exhibits on a variety of topics, some of which I developed. Look in the educator's resource area.
http://observe.ivv.nasa.gov

Yohkoh Public Outreach Project - Sun/Earth connection materials for K-12 classrooms. Great source for solar movies.
http://solar.physics.montana.edu/YPOP/

NASA Center for Educational Resources Project - NASA materials for astronomy education
http://btc.montana.edu/ceres/

Real-Time Science Data Access for Classroom Investigation. Great source of science data and images. Not limited to astronomy. Has strong earth science emphasis as well.
URL: http://www.math.montana.edu/~tslater/real-time/

The Astronomy Café by Sten Odenwald
Great site for introduction to astronomy with many movies, images, questions answered, astronomy artcles, etc. Site is also available on CD-ROM
http://itss.raytheon.com/cafe/

American Astronomical Society Education Resources. Extensive resources on-line, and growing!
http://www.aas.org/~education/resource.html

Astronomy Educational Resources at the Institute for Astronomy at the University of Hawaii. A superior resource.
http://www.ifa.hawaii.edu/tops/resources.html

Skyview
SkyView is a Virtual Observatory on the Net generating images of any part of the sky at wavelengths in all regimes from Radio to Gamma-Ray.
http://skyview.gsfc.nasa.gov/

Astronomical pictures & animations
An online 4.8 gigabytes astronomical server
http://graffiti.u-bordeaux.fr/MAPBX/roussel/astro.english.html

Practical Uses of Math and Science (PUMAS)
The on-line journal of math and science examples for pre-college education
http://pumas.jpl.nasa.gov

Stanford Solar Center
Large site devoted to the Sun
http://solar-center.stanford.edu/

WebLinks from Alan Gould at Lawarence Hall of Science
Alan keeps his astronomy education and related bookmarks on the web where they are available to science educators worldwide. Great links for students and teachers.
http://www.lhs.berkeley.edu/SII/SIImorelinks.html

His bookmarks include the following area:

- NASA sites
- Latest Hubble Space Telescope images
- Ecology and Energy
- Companies
- Education
- Astronomy Places
- Earth Science
- Miscellaneous Places

Some Resources for College Science Teaching

The Journal of College Science Teaching (JCST)
http://www.nsta.org/jcst/

The Journal of College Science Teaching (JCST) is a refereed journal published by the National Science Teachers Association (NSTA) for an audience of college and university teachers of introductory and advanced science courses. JCST communicates innovative, effective techniques to improve interdisciplinary teaching strategies for instructing both science majors and nonmajors.

In addition to publishing an average of six major feature articles per issue, JCST contains a timely science-news section, book and instructional media reviews, articles describing scientifically historic sites around the world, exciting laboratory demonstrations, opinion pieces, and explorations of the relationship between science and society.

Some Different Approaches to Teaching Physics:
Descriptions of Sample Physics Classes from Innovators
http://www.physics.umd.edu/~inted/clssdscr.htm

University of California, Berkeley Teaching Suggestions
The 212 suggestions presented in this compendium were, with few exceptions, contributed by members of the faculty were gathered from recipients of the Distinguished Teaching Award given annually by the Academic Senate Committee on Teaching.
http://uga.berkeley.edu/sled/compendium/

Excellent Review Paper: Cognitive Aspects of Learning and Teaching Science
by Jose P. Mestre, Department of Physics & Astronomy
University of Massachusetts, Amherst, MA 01003-4525 USA
http://www-perg.phast.umass.edu/papers/ReviewPaper/ReviewPaper.html

Readings on Pedagogy
Web readings on the subjects of cognition and learning styles, constructivist inquiry, learning communities, simulation and other topics.
http://seamonkey.ed.asu.edu/~vito/mec/readings.html

Pathways to School Improvement
This site addresses directly addresses critical school improvement issues including content, environment, educators, students, and teaching. Although it is aimed at the pre-college level, it summarizes well the current trends in education.
http://www.ncrel.org/sdrs/topics.htm

Astronomy Organizations and Catalog Supply Houses

American Association of Variable Star Observers (AAVSO), 25 Birch Street, Cambridge, MA 02138. Amateur and professionals interested in variable stars and variable star observing.

American Astronomical Society (AAS), Education Office, Organizes workshops for local teachers at AAS meetings, has produced an excellent brochure on as astronomy as a career, and coordinates astronomy education activities among professional astronomers.

Astronomical Society of the Pacific, 390 Ashton Avenue, San Francisco CA 94112, (415) 337-1100. Great source of videotapes, slides, videodisks, and other educational resource materials.

Canadian Astronomical Society, c/o David Dunlap Observatory, Box 360, Richmond Hill, ON L4C 4Y6 Canada. Active in astronomy education.

Edmund Scientific, 101 East Gloucester Pike, Barrington, NJ 08007-1380

International Dark-Sky Association, 3225 N. First Ave. Tucson, AZ 85719 USA. Dedicated to preserving the sky as a resource to be cherished and to fighting light pollution. Has created an array of educational materials and slide sets.

Kalmbach Publishing, 21027 Crossroads Circle, P.O. Box 1612, Waukesha, WI 53187. Publisher of *Astronomy* magazine.

MMI Corporation, P. O. Box 19907, Baltimore MD 21211, Extensive catalog of astronomy and earth science supplies including many hard to find items for teachers. Extensive collection of software, globes, videos, slides.

National Space Society, 600 Pennsylvania Avenue, S.E. Suite 201, Washington, DC 20003. An advocacy organization for manned space exploration. Publishes a magazine, newsletter, and catalog.

Orion Telescopes and Binoculars, P. O. Box 1815, Santa Cruz, CA 95061-1815

The Planetary Society, 65 N. Catalina Ave., Pasadena CA 91106-2301. Organization that has been active in advocating greater efforts towards planetary exploration. Publishes *The Planetary Report*.

Royal Astronomical Society of Canada (RASC), 136 Dupont St. Toronto, ON M5R 1V2 Canada. Very active society of amateur and professional astronomers. The RASC Observer's Handbook is a very useful observing and teaching guide.

***Sky Publishing*, Corp**, 49 Bay State Rd., Cambridge MA 02138. Publisher of *Sky and Telescope* . Has an extensive catalog of astronomy materials, including books, posters, software, globes, and observing supplies.

Willman-Bell, Inc., P.O. Box 35025, Richmond, Virginia 23235. Publisher of many astronomy books. Excellent extensive catalog of books, software, and mirror polishing supplies.

International Dark-Sky Association

One of the most important organization to support is the International Dark-Sky Association (IDA). IDA is a 501(c)(3) membership based organization dedicated to preserving the beauty of our night skies by promoting quality outdoor lighting practices and by educating the public on the benefits of preserving our awareness of the stars. IDA pursues goals that are mutually beneficial to astronomy and society. IDA's philosophy is that the energy consumed to produce night lighting is used most economically and effectively if the light stays on the ground where it is needed, is not wastefully scattered into the night sky, and eliminates dangerous glare. Quality outdoor lighting conserves energy and improves safety and security by using only the amount of light needed, where it is needed, and when it is needed.

IDA has accumulated a membership of nearly 4000 spanning the United States and 71 foreign countries. About 200 are organizational members, such as observatories, universities, astronomy clubs, lighting companies, and cities. IDA benefits from the support of professional and public communities. Members have effected lighting code changes that preserve safety of constituents while reducing light pollution.

The population of the world has reached six billion individuals. Our cities and suburban areas grow without consumer input. In the face of growth, thoughtful homeowners consider the effect of sprawl on their environment and lifestyles. Residents who seek to protect their quality of life turn to advocacy organizations to guide them in methods and means to reach their goals. There is a place for organizations such as IDA to be a resource for consumers and homeowners. Light pollution is an growing problem, threatening both astronomical facilities and the dark sky that is part of our human heritage.

IDA currently offers education tools for the public in the form of quarterly newsletters, more than 150 information sheets, several slide sets, videos, and similar resources in support of its goals. Specific educational resources include brochures, economic information, lighting codes, examples of lighting designs and good and bad lighting, references, speakers, a list-server, individual staff responses to inquiries, and annual and regional meetings. IDA's web page is as a resource for the media as well as many other readers.

IDA has three broad goals for the future: to expand its technical expertise; to enhance education outreach to both the public and schools; and to reinforce the organization's position as *the* resource for appropriate lighting codes and information.

Contact Information:

International Dark-Sky Association (IDA)
3225 N. First Avenue
Tucson, AZ 85719-2103
(877) 600-5888 (toll free for membership)
ida@darksky.org (IDA office)

PASS Project (Planetarium Activities for Student Success)

Cary Sneider, Alan Friedman and Alan Gould, Editors

Designed for both experienced planetarium instructors and teachers using a planetarium for the first time, this series provides a wealth of ready-to-go ideas and practical suggestions for planning and presenting entertaining and educationally effective programs for students. The first four books in the series provide a general orientation to astronomy and space science education, with applications for both the planetarium and classroom settings. The other eight books each present a complete planetarium program and related classroom activities. The books are available for purchase individually or as a full set. For more information about PASS, call the Lawrence Hall of Science at (510) 642-5863.

The PASS Series Includes:

1. Planetarium Educator's Workshop Guide
2. Planetarium Activities for Schools
3. Resources for Teaching Astronomy & Space Science
4. A Manual for Using Portable Planetariums
5. Constellations Tonight
6. Red Planet Mars
7. Moons of the Solar System
8. Colors From Space
9. How Big Is the Universe?
10. Who "Discovered" America?
11. Astronomy of the Americas
12. Stonehenge

Audience: Teachers and Planetarium Instructors
Age/Grade Level: Grades K-12
Subject Area: Astronomy and Space Science
Format: Books, 30-70 pages each, black and white photos, illustrations
Price: $11.95 per book; $132.00 per set of twelve
Item Number for Set of Twelve: AST300

Sky Challenger

Much more than the standard star-finder wheel, Sky Challenger is a unique tool for learning astronomy. With six interchangeable wheels full of activities, this kit makes learning enjoyable and accessible. The following wheels comprise the Sky Challenger: Introductory Wheel, Native American Constellations, Binocular Sky Treasure Hunt, Test Your Eyes-Test the Skies, Where Are the Planets?, and Invent Your Own Constellations. Each Sky Challenger also includes a Star Clock, for telling time based on the position of well-known constellations; and a copy of the Star Gazer's Gazette, containing articles about astronomy and tips for using your Sky Challenger. The Sky Challenger can be used anywhere in the continental United States. For more information call 510-642-5863.

Audience: Children and adults
Age/Grade Level: Age 8 and up
Subject Area: Astronomy
Format: Printed materials kit, 10"-diameter wheels
Price: $8.95
Item Number: AST110

Case of 20: $163.00
Item Number: AST111

Architect Sun Angle Calculator

There are cardboard "calculators" used by architects which allow you to predict the Sun's position. The best known example was made by Pilkington (a glass company) and is available for $10 (check only)

Ms. Kathy Raszka
Pilkington
811 Madison Ave
Toledo Ohio 43697
Phone (419) 247-3731

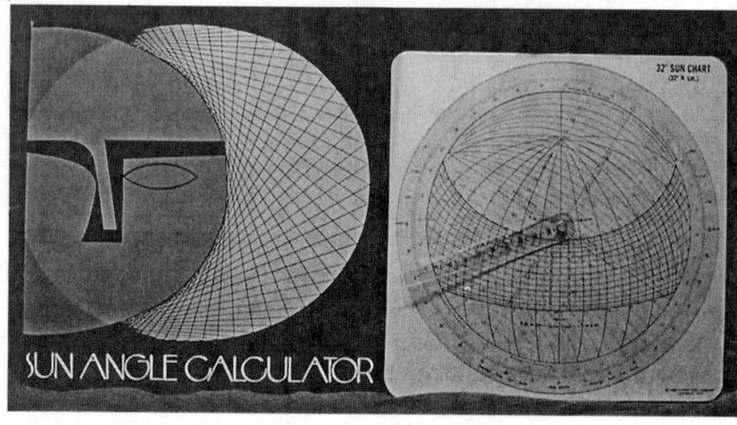

Scholarship Opportunities

Intel Annual Science Talent Search
Provides scholarships for top science students each year. The top award this year was a 100,000 scholarship. Winners are selected from the entrants, based on a written report of an independent science, math or engineering research project. Application materials are usually available mid-August each year, with a deadline of December 1.

Intel Science Talent Search
Science Service
1719 N Street, NW
Washington, DC, 20036
http://www.sciserv.org/sts

The National Space Science Data Center

The NSSDC provides access to a wide variety of astrophysics, space physics, solar physics, lunar and planetary data from NASA space flight missions, in addition to selected other data and some models and software. NSSDC provides access to online information bases about NASA and non-NASA data at the NSSDC and elsewhere as well as the spacecraft and experiments that have or will provide public access data.

The NSSDC also provides an extensive catalog of CD-ROMs and have a great number of planetary and other images on-line. One of their most popular planetary image CD-ROMs is only $10. The CD-ROMS which contain data and archival images are particularly useful for class projects for advanced students.

Astronomical Software

The amount of software available for astronomy has become too large to discuss in detail. Please see the following web page for an excellent introduction to astronomical software. It is the software review page for John Mosley at Griffith Observatory, Los Angeles.
http://www.GriffithObs.org/software.html

The Sky and Telescope web site has a listing of downloadable astronomy programs. This includes some BASIC programs as well as a listing of free or inexpensive astronomy programs. It also includes a listing of software vendors.
http://www.skypub.com/resources/software/software.shtml

An extensive listing of planetarium/sky viewing software is available at
http://seds.lpl.arizona.edu/nineplanets/astrosoftware.html

Educational Newsletters and Publications

There are several newsletters available for educators which present recent scientific discoveries at a level designed to be used in the classroom. Below are the addresses and contacts for the various newsletters.

Name	Description	Price	Address
Universe in the Classroom	Newsletter on Teaching Astronomy (contains educational activities & latest news items in astronomy)	free	Teachers' Newsletter Dept. Astron. Soc. of the Pacific 390 Ashton Ave. San Francisco, CA 94122

http://www.aspsky.org/education/tnl.html

Note: The newsletter is available in Spanish, and has been translated by volunteers into more than a dozen other languages! This newsletter is available free of charge to teachers, school librarians and administrators, and youth-group leaders who request it on institutional stationery. Highly recommended for all astronomy educators. Full of astronomy teaching ideas and demonstrations and analogies.

Planetary Report	Newsletter for Planetary Society-current events, news, etc.	$25/yr The Planetary Society 65 N. Catalina Avenue (626) 793-5100 email:tps@mars.planetary.org

The Planetary Society also distributes the Mars Underground News and the Bioastronomy News.

StarDate

More a small magazine than a newsletter. A publication of the University of Texas, McDonald Observatory. Excellent quality and very useful for amateurs and educators alike. Get a subscription today. The same group hosts a wonderful radio show, available also in Spanish. You can explore all of the aspects of their programs at StarDate Online
http://stardate.utexas.edu/

They also have an excellent on-line Astronomical Glossary at
http://stardate.utexas.edu/resources/astroglossary/glossary_A.html

Public Information Office of Space Telescope Science Institute

Web address:
http://oposite.stsci.edu/

This is a great resource for space astronomy images and movies. The site has many well tested education resources.
At the site you can find:
1. What's New
2. News & Views
    - Latest News from Hubble
    - Hubble Press Release Archive
    - Servicing Mission 3A
3. Gallery
    - Hubble Pictures!
    - Cool Movies & Animations
    - Greatest Hits 1990-1998

Greatest Hits 1990-1995
Gallery of Planetary Nebulae

4. Education
   Amazing Space- K-12 interactive activities on the Web
   Education at Hubble
   Origins- searching for our cosmic roots
5. Sci•Tech
   The Story of the Hubble Space Telescope
   STScI at a Glance
   About the Hubble Space Telescope
   Behind the Pictures
   Tour the Cosmos- a multimedia experience

---Some Curriculum Projects to Watch---

Life in the Universe Science Curriculum Development Project

From the SETI website:
"An educational project conducted by the SETI Institute with sponsorship from the National Science Foundation and NASA. The "Life in the Universe" project is creating supplementary science curricula for elementary and middle school students. The intent of the project is to take advantage of the intense interest that young students have in the existence of extraterrestrial life to motivate their studies of science and other subjects. The curriculum activities are drawn from the experience of SETI scientists, whose research requires considerations of astronomy, life sciences, earth sciences, chemistry, physics, mathematics, engineering, and many other disciplines. The lessons, activities, and experiments now under development teach standard science concepts in the context of questions about the nature of life, its possible presence elsewhere in the universe, and the physical/chemical settings provided by space, planets, and stars, and within which other life might be sought. "

"The science activities are designed by teachers and tested in their classrooms. They are "hands-on" in approach and philosophy, and encourage students to express their learning in alternative ways. The activities also encourage questioning, and provide many opportunities to reinforce critical thinking skills. Initial classroom trials demonstrate a high degree of student and teacher acceptance, and confirm the appeal of these activities to students of diverse backgrounds."

The goals of the SETI Science Curriculum Project include:

--Improving the teaching of science in elementary and middle schools
--Using the interdisciplinary nature of SETI to create natural links between various topics
--Making science a tangible and enjoyable subject for students.
Six Teachers' Guides for grades 3 - 9 that contain "hands-on" science activities have been created (all guides are available):

Grades 3-4, one guide: "The Science Detectives"
Grades 5-6, three guides,
"Evolution of a Planetary System,"
"How Might Life Evolve on Other Worlds?" and
"The Rise of Intelligence and Culture."
Middle school, two guides:
"Life: Here? There? Elsewhere?"
"Project Haystack"

A Sample Lesson from each guide is available. Components of the "Life in the Universe" curriculum include:
--Teachers' Guides, described above. These include black-line masters to make student work books.
--Specially commissioned artwork and transparencies, depicting Earth's geological and biological history
--Cards and posters
--Videos for two of the guides

The materials are available from the Astronomical Society of the Pacific and Libraries Unlimited/Teacher Ideas Press
Dept. 9503
PO Box 6633
Englewood, Colorado
Phone: 800-237-6124
Fax: 303-220-8843

Astronomy Village: Investigating the Universe

The "Astronomy Village: Investigating the Universe" package was developed by the NASA Classroom of the Future Program at Wheeling Jesuit University. It is designed as a supplement to the 9th grade earth science or physical science curriculum, although it has been tested and used at a variety of grade levels. The basis of the software is that students research teams go the "The Astronomy Village" to work on an astronomical investigation for about three weeks. In the Astronomy Village, students actively construct inferences, relations, comparisons, questions, mental models, and the like. They actively solve problems and use image processing. There is a strong mathematics component as well.

In this CD-ROM based multimedia program, student teams can pursue one of ten research "investigations". In each investigation they are guided by a mentor, receive email, hear a lecture in the Village auditorium, and make observations using ground or space-based telescopes in a virtual observatory. The students also process data using the *NIH Image* image processing program. They also keep a detailed logbook of their research activities. They can run simulations on stellar evolution and can manipulate 3-D visualization tools. At the end, they present their research to others in their class and answer questions in a press conference about their results.

The investigations emphasize open-ended research topics that are similar to front-line astronomy research questions. For example, the investigations revolve around searches for: obscured supernovae using neutrino data; Earth-crossing asteroids; planets around nearby stars; undiscovered nearby stars; and protoplanetary disks. Other investigations address long standing astronomical problems: characterizing galaxy clusters; comparing views towards the Milky Way with views away from it; using Cepheids for distance determination; characterizing how star forming regions look at different wavelengths; and astronomical site selection.

The richness of the Astronomy Village environment makes it extremely valuable for informal education. The simulation tools, videos, lectures, library articles, and image browser make it a self-enclosed astronomy resource that can be used for brief periods of time, or for hours, in a home or science center setting.

The program is available in Mac format and is being converted to PC format as this book goes to press (Spring, 2000). For more information contact the Editor, who was a project member or go to the Center for Educational Technologies at Wheeling Jesuit University web site at www.cet.edu.

Voyages Through Time

This project provides an integrated, technology-based, high school science curriculum.

Evolutionary change is a powerful framework for studying our world and our place therein. It is a recurring theme in every realm of science: over time, the universe, the planet Earth, life, and human technologies all change, albeit on vastly different scales. Evolution offers scientific explanations for the age-old question, "Where did we come from?" In addition, historical perspectives of science show how our understanding has evolved over time. The complexities of all of these systems will never reveal a "finished" story. But it is a story of epic size, capable of inspiring awe and of expanding our sense of time and place, and eminently worthy of investigating. This story is the basis of *Voyages Through Time*.

The SETI Institute, the California Academy of Sciences, NASA Ames Research Center, and San Francisco State University are developing standards-based curriculum materials for a one-year high school integrated science course centered on the unifying theme of evolution. Scientists, teachers, curriculum writers, and media specialists have created six modules that integrate astronomical, geological, and biological sciences. The sequence of lessons in each module is designed to promote students' understanding and skills as defined by the *National Science Education Standards* and *Benchmarks for Science Literacy*.

The six week modules, Cosmic Evolution, Planetary Evolution, Origin of Life, Diversification of Life, Hominid Evolution, and Evolution of Technology, use the constructivist approach of engage, explore, explain, elaborate, and evaluate (BSCS, 1996) as an instructional framework. The core lessons for all six modules is provided on a teacher's CD-ROM; instructional guidelines, science background information, resources (print, audiovisual, software, WWW sites), as well as student handouts will be included on the CD. Each module also has a soft-cover reader of articles from popular science magazines and a student's CD-ROM of databases and images. These products will be published as a complete set for use as a year-long integrated science course and will also be available as individual modules for use in discipline-based courses.

The content and instructional approach of the *Voyages Through Time* curriculum addresses national standards and recommendations (NRC, AAAS, NSTA), all of which acknowledge evolution as an overarching theme and call for activity-based learning for students. The curriculum also meets the need for quality science materials that incorporate technology, while accommodating schools still building their technology capacities and promoting students' basic literacy.

Teachers from the San Francisco Bay Area and scientists from the collaborating institutions and members of our science advisory board work with the project leadership, curriculum writers, and multimedia specialists to design the curriculum. The first three Voyages modules are being pilot tested in local schools and field-tested in schools across the nation. Reviews of content for pedagogy and science accuracy continue throughout the development process, with the goal of producing final materials of the highest quality.

For further information, contact

Edna DeVore
Director of Educational Programs
SETI Institute
2035 Landings Dr.
Mountain View, CA 94043
edevore@seti.org

650-961-6633

http://www.seti.org

Astronomy Village: Investigating the Solar System

Funded by a grant from the National Science Foundation, Astronomy Village: Investigating the Solar System is a CD-ROM-based multimedia package introducing solar system astronomy. This innovative package is intended to complement and extend the science curricula for grades 5 through 7. Students will be actively engaged in practical explorations of some of the key questions in solar system astronomy. The package is philosophically similar to the *Astronomy Village: Investigating the Universe*, the award-winning CD-ROM targeting the high school level.

The Astronomy Village: Investigating the Solar System CD-ROM was developed by a team led by Dr. Steven Croft and Dr. Steven McGee, both with the Center for Educational Technologies at Wheeling Jesuit University, and Dr. Stephen Pompea of Pompea and Associates. The goal of *Astronomy Village: Investigating the Solar System* is to provide middle school teachers and students with a multimedia instructional package that is aligned with the tenets of the National Science Education Standards. The program incorporates recent findings in effective teaching and learning practices and provides classrooms with extensive resources on solar and planetary astronomy. With the guidance of their teachers, students engage in genuine scientific inquiry and investigation in a rich visualization and modeling environment.

Students using *Astronomy Village* will be motivated to learn key concepts in astronomy, engage in scientific inquiry, explore current solar system research, and make use of technology to explore and analyze data.

For more information, see http://www.cet.edu/av2
or, contact Dr. Steven Croft at:
Center for Educational Technologies
Wheeling Jesuit University
316 Washington Avenue
Wheeling, WV 26003
304/243-2491 (voice) 304/243-2497 (fax)
scroft@cet.edu

Image Processing for Teaching

The Image Processing for Teaching project, developed at the University of Arizona under the leadership of Professor Richard Greenberg, has proven to be an effective way to excite students about science and mathematics, and to offer extensive opportunities for exploration and discovery. Since 1990 teachers have been trained in the use of digital image processing and in the scientific content of digital images. Experience at test sites shows that is an effective way to excite both traditional and nontraditional learners, male and female, about science and mathematics. Curriculum activities have been developed in a wide range of subject areas, including astronomy. The Center for Image Processing in Education was established to meet the widespread demand from schools, faculty, and scientists for national dissemination of image processing curriculum materials. This center is coordinating the efforts of participating scientists and expert teachers to support science and mathematics education programs. For additional information contact:

Center for Image Processing in Education
P.O. Box 13750
Tucson, AZ 85732-3750
Phone: 800/322-9884
www.cipe.com

MicroObservatory

The Harvard-Smithsonian Center for Astrophysics, with funds from the NSF, has developed a network of portable, automated, CCD-based reflecting telescopes accessible to secondary students at their schools and over Internet. The basic goal is a low-cost, computer-driven, automated imaging telescope that can be used by students for research. Data can be collected remotely, transferred over Internet, and analyzed. The telescope control software allows the user to choose resolution, exposure time, and filters. The user views their object by using a computer monitor, just as most researchers in astronomy do today. The MicroObservatory Project currently has a network of five automated telescopes that can be controlled over the Internet. The MicroObservatory software has won a number of awards as well.

For more information see
http://mo-www.harvard.edu/MicroObservatory/

Telscopes in Education

The Telescopes in Education (TIE) program brings the opportunity to use a remotely controlled telescope and charge-coupled device (CCD) camera in a real-time, hands-on, interactive environment to students around the world. TIE enables students to increase their knowledge of astronomy, astrophysics, and mathematics; improve their computer literacy; and strengthen their critical thinking skills. Telescopes In Education is a Mount Wilson Institute program sponsored by the National Aeronautics and Space Administration (NASA) and developed through the efforts of numerous volunteers, businesses, and supporting organizations including the Jet Propulsion Laboratory (JPL) of the California Institute of Technology (Caltech).

The TIE program currently utilizes a science-grade 24-inch reflecting telescope located at the Mount Wilson Observatory, high above the Los Angeles basin in the San Gabriel Mountains of Southern California. The telescope has been used by students in grades K-12 to observe galaxies, nebulae, variable stars, eclipsing binaries, and in other ambitious projects and experiments. Hundreds of schools in the US and around the world (including Australia, Canada, England, and Japan) have successfully used the prototype telescope on Mount Wilson. Through TIE, students have rediscovered and cataloged a variable star and assisted the Pluto Express project at NASA's Jet Propulsion Laboratory to revise the ephemeris (orbital location) for the planet Pluto. For more information see the TIE website at:

http://tie.jpl.nasa.gov/tie/

Hands-On Universe

For the past seven years, with support from the National Science Foundation, the Department of Defense, and the Department of Energy, the Hands-On Universe (HOU) project at the University of California, Berkeley has developed and pilot tested an educational program that enables high school students to request their own observations from professional observatories. Students download telescope images to their classroom computers and use the powerful HOU image processing software to visualize and analyze their data. High school curriculum developed by HOU integrates many of the science and math topics and skills outlined in national standards into open- ended astronomical investigations.

HOU is also developing activities and tools for middle school students and products for informal science education centers. HOU materials can be found in more than 500 classrooms in regional high school networks across the world, including U.S. Department of Defense Dependent Schools. HOU is a collaboration of Lawrence Hall of Science, TERC Inc. (Cambridge, Massachusetts), Adler Planetarium (Chicago), and Yerkes Observatory (Williams Bay, Wisconsin), as well as a network of educators and astronomers from all over the world. More information can be found at:

http://hou.lbl.gov/

Appendix 1: Astronomy Diagnostic Test (ADT) Information

The development of this diagnostic test, titled Introductory Astronomy Survey, was accomplished the multi-institutional Collaboration for Astronomy Education Research (CAER) including, among many others, Jeff Adams, Rebecca Lindell Adrian, Christine Brick, Gina Brissenden, Grace Deming, Beth Hufnagel, Tim Slater, and Michael Zeilik.

The test is free for introductory astronomy faculty members to use, with some agreed guidelines. The test is not to be altered in any way, questions are not to be left out, questions are not to be removed and placed on course quizzes or exams, and the test items are not to be distributed to students lest the items find their way into "student files." In exchange for using the ADT, the authors respectfully request that you submit scores to the national database (please e-mail Beth Hufnagel for details).

If you would like a copy of the "answer key," please email Beth Hufnagel at hufnagel@astro.umd.edu with your name, institution, and what classes are taught.

The first 21 questions represent the content portion of the test while the final 12 questions collect demographic information. The demographic information would be helpful to us but is not crucial.

The test is available on the web site as an Adobe Acrobat file and the test is also available in a Spanish version. If you would like to compare your class with others, please feel free to download the COMPARISON DATABASE Excel spreadsheet (pretest scores). If you can not find a comparative class, please email Beth Hufnagel with a description of your class and she will help you.

More details are available at http://solar.physics.montana.edu/aae/adt

Introductory Astronomy Survey

1. As seen from your current location, when will an upright flagpole cast no shadow because the Sun is directly above the flagpole?
   A. Every day at noon.
   B. Only on the first day of summer.
   C. Only on the first day of winter.
   D. On both the first days of spring and fall.
   E. Never from your current location.

2. When the Moon appears to completely cover the Sun (an eclipse), the Moon must be at which phase?
   A. Full
   B. New
   C. First quarter
   D. Last quarter
   E. At no particular phase

3. Imagine that you are building a scale model of the Earth and the Moon. You are going to use a 12-inch basketball to represent the Earth and a 3-inch tennis ball to represent the Moon. To maintain the proper distance scale, about how far from the surface of the basketball should the tennis ball be placed?
   A. 4 inches (1/3 foot)
   B. 6 inches (1/2 foot)
   C. 36 inches (3 feet)
   D. 30 feet
   E. 300 feet

4. You have two balls of equal size and smoothness, and you can ignore air resistance. One is heavy, the other much lighter. You hold one in each hand at the same height above the ground. You release them at the same time. What will happen?
   A. The heavier one will hit the ground first.
   B. They will hit the ground at the same time.
   C. The lighter one will hit the ground first.

5. How does the speed of radio waves compare to the speed of visible light?
   A. Radio waves are much slower.
   B. They both travel at the same speed.
   C. Radio waves are much faster.

6. Astronauts inside the Space Shuttle float around as it orbits the Earth because
   A. there is no gravity in space.
   B. they are falling in the same way as the Space Shuttle.
   C. they are above the Earth's atmosphere.
   D. there is less gravity inside the Space Shuttle.
   E. more than one of the above.

7. Imagine that the Earth's orbit were changed to be a perfect circle about the Sun so that the distance to the Sun never changed. How would this affect the seasons?
   A. We would no longer experience a difference between the seasons.
   B. We would still experience seasons, but the difference would be much LESS noticeable.
   C. We would still experience seasons, but the difference would be much MORE noticeable.
   D. We would continue to experience seasons in the same way we do now.

8. Where does the Sun's energy come from?
   A. The combining of light elements into heavier elements
   B. The breaking apart of heavy elements into lighter ones
   C. The glow from molten rocks
   D. Heat left over from the Big Bang

9. On about September 22, the Sun sets directly to the west as shown on the diagram below. Where would the Sun appear to set two weeks later?
   A. Farther south B. In the same place C. Farther north

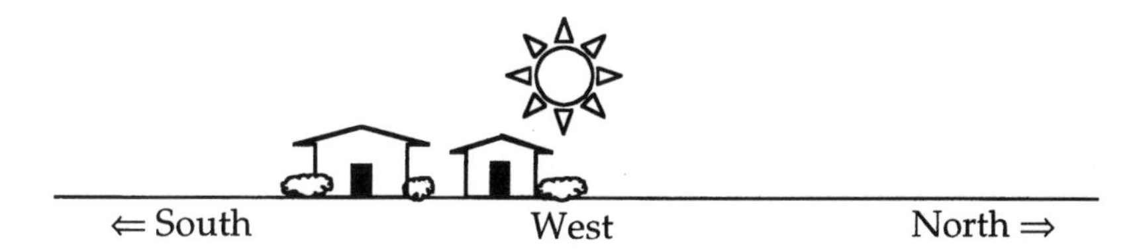

10. If you could see stars during the day, this is what the sky would look like at noon on a given day. The Sun is near the stars of the constellation Gemini. Near which constellation would you expect the Sun to be located at sunset?
    A. Leo
    B. Cancer
    C. Gemini
    D. Taurus
    E. Pisces

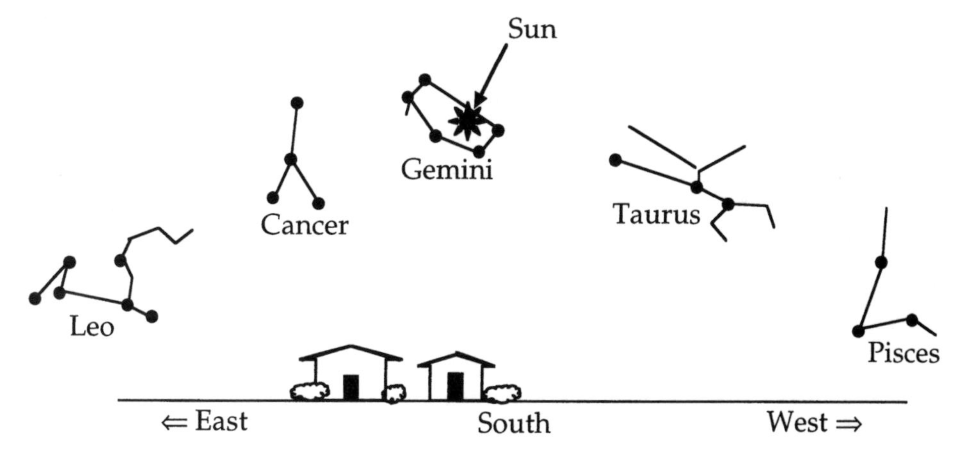

11. Compared to the distance to the Moon, how far away is the Space Shuttle (when in space) from the Earth?
    A. Very close to the Earth
    B. About half way to the Moon
    C. Very close to the Moon
    D. About twice as far as the Moon

12. As viewed from our location, the stars of the Big Dipper can be connected with imaginary lines to form the shape of a pot with a curved handle. To where would you have to travel to first observe a considerable change in the shape formed by these stars?
    A. Across the country
    B. A distant star
    C. Europe
    D. Moon
    E. Pluto

13. Which of the following lists is correctly arranged in order of closest-to-most-distant from the Earth?
    A. Stars, Moon, Sun, Pluto
    B. Sun, Moon, Pluto, stars
    C. Moon, Sun, Pluto, stars
    D. Moon, Sun, stars, Pluto
    E. Moon, Pluto, Sun, stars

14. Which of the following would make you weigh half as much as you do right now?
    A. Take away half of the Earth's atmosphere.
    B. Double the distance between the Sun and the Earth.
    C. Make the Earth spin half as fast.
    D. Take away half of the Earth's mass.
    E. More than one of the above

15. A person is reading a newspaper while standing 5 feet away from a table that has on it an unshaded 100-watt light bulb. Imagine that the table were moved to a distance of 10 feet. How many light bulbs in total would have to be placed on the table to light up the newspaper to the same amount of brightness as before?
    A. One bulb.
    B. Two bulbs.
    C. Three bulbs.
    D. Four bulbs.
    E. More than four bulbs.

16. According to modern ideas and observations, what can be said about the location of the center of the Universe?
    A. The Earth is at the center.
    B. The Sun is at the center.
    C. The Milky Way Galaxy is at the center.
    D. An unknown, distant galaxy is at the center.
    E. The Universe does not have a center.

17. The hottest stars are what color?
    A. Blue
    B. Orange
    C. Red
    D. White
    E. Yellow

18. The diagram below shows the Earth and Sun as well as five different possible positions for the Moon. Which position of the Moon would cause it to appear like the picture at right when viewed from Earth?

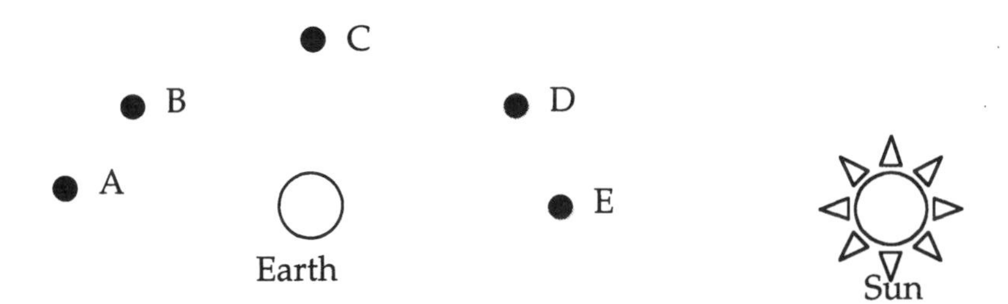

19. You observe a full Moon rising in the east. How will it appear in six hours?

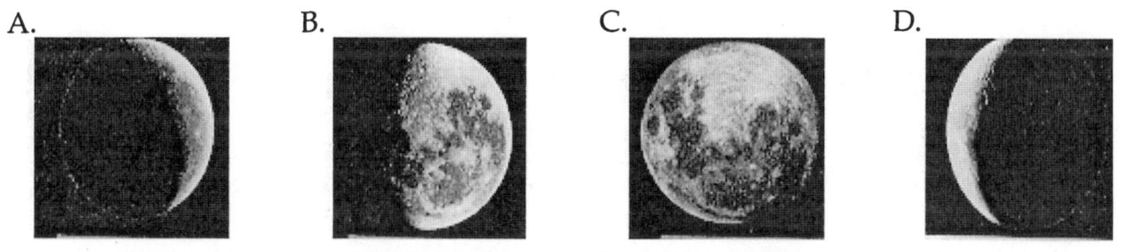

Images © UC Regents/Lick Observatory. Unauthorized use prohibited.

20. With your arm held straight, your thumb is just wide enough to cover up the Sun. If you were on Saturn, which is 10 times farther from the Sun than the Earth is, what object could you use to just cover up the Sun?
    A. Your wrist
    B. Your thumb
    C. A pencil
    D. A strand of spaghetti
    E. A hair

21. Global warming is thought to be caused by the
    A. destruction of the ozone layer.
    B. trapping of heat by nitrogen.
    C. addition of carbon dioxide.

22. In general, how confident are you that your answers to this survey are correct?
    A. Not at all confident (just guessing)
    B. Not very confident
    C. Not sure
    D. Confident
    E. Very confident

23. What is your college major (or current area of interest if undecided)?
    A. Business
    B. Education
    C. Humanities, Social Sciences, or the Arts
    D. Science, Engineering, or Architecture
    E. Other

24. What was the last math class you completed prior to taking this course?
    A. Algebra
    B. Trigonometry
    C. Geometry
    D. Pre-Calculus
    E. Calculus

25. What is your age?
    A. 0-20 years old
    B. 21-23 years old
    C. 24-30 years old
    D. 31 or older
    E. Decline to answer

26. Which best describes your home community (where you attended high school)?
    A. Rural
    B. Small town
    C. Suburban
    D. Urban
    E. Not in the USA

27. What is your gender?
    A. Female
    B. Male
    C. Decline to answer

28. Which best describes your ethnic background?
    A. African-American
    B. Asian-American
    C. Native-American
    D. Hispanic-American
    E. None of the above (see question 29 below)

29. Which best describes your ethnic background?
    A. African (not American)
    B. Asian (not American)
    C. White, non-Hispanic
    D. Multicultural
    E. None of the above (see question 28 above)

30. How good at math are you?
    A. Very poor
    B. Poor
    C. Average
    D. Good
    E. Very good

31. How good at science are you?
    A. Very poor
    B. Poor
    C. Average
    D. Good
    E. Very good

32. Which best describes the level of difficulty you expect/experienced from this course?
    A. Extremely difficult for me
    B. Difficult for me
    C. Unsure
    D. Easy for me
    E. Very easy for me

33. How many astronomy courses at the college level have you taken?
    A. I'm re-taking this course.
    B. This is my first college-level astronomy course.
    C. This is my second college-level astronomy course.
    D. I've completed more than two other college-level astronomy courses.

End of Survey
